Stability of Life on Earth

Principal Subject of Scientific Research in the 21st Century

Springer

Berlin
Heidelberg
New York
Hong Kong
London
Milan
Paris
Tokyo

Kirill Ya. Kondratyev, Kim S. Losev,
Maria D. Ananicheva and Irina V. Chesnokova

Stability of Life on Earth

Principal Subject of Scientific Research in the 21st Century

Springer

Published in association with

Praxis Publishing
Chichester, UK

Professor K. Ya. Kondratyev
Counsellor of the Russian Academy of Sciences
Scientific Research Centre for Ecological Safety
Nansen Foundation for Environment
  and Remote Sensing
St Petersburg
Russia

Professor Dr K. S. Losev
VINITI
Moscow University
Russia

Dr M. D. Ananicheva
Institute of Geography
Moscow
Russia

Dr I. V. Chesnokova
Institute of Lithosphere
Moscow
Russia

SPRINGER–PRAXIS BOOKS IN ENVIRONMENTAL SCIENCES
SUBJECT *ADVISORY EDITOR*: John Mason B.Sc., M.Sc., Ph.D.

ISBN 3-540-20328-1 Springer-Verlag Berlin Heidelberg New York

Springer-Verlag is a part of Springer Science+Business Media (springeronline.com)

Bibliographic information published by Die Deutsche Bibliothek

Die Deutsche Bibliothek lists this publication in the Deutsche Nationalbibliografie;
detailed bibliographic data are available from the Internet at http://dnb.ddb.de

Library of Congress Cataloging-in-Publication Data
    Stability of life on earth : principal subject of scientific research in the 21st century / Kirill
    Ya Kondratyev ... [et al.].
      p. cm.
    Includes bibliographical references (p.)
    ISBN 3-540-20328-1 (alk. paper)
    1. Biotic communities.   2. Nature—Effect of human beings on.   3. Global environmental
    change.   4. Global warming.   5. Biogeochemical cycles.   I. Kondrat'ev, K. IA. (Kirill
    IAkovlevich)
    QH541.S654 2003
    577.27—dc22   2003065299

Cover design: Jim Wilkie
Project Management: Originator Publishing Services, Gt Yarmouth, Norfolk, UK

Printed on acid-free paper

Contents

Preface

The purpose of the book is to show that life – which consists of the total number of natural organisms, or the Earth biota, and is organized in mutually dependent communities – is a truly self-organizing system that forms, regulates and ensures the dynamic stability of the environment. The primary regulating elements are elementary structural units called "facies" in Landscape Science and "biogeocenoses" in Biology and Ecology. They are the elementary units of the biological cycle (Chapters 1–3) that determine the mechanism of regulation and stabilization of the environment.

Humankind, like natural biota, is also building its environment – the "technosphere" – through continuous destruction of the elementary structural units of the biota (i.e., destruction of wildlife) as can be amply illustrated throughout the history of the development of civilization. Increasingly sophisticated technologies are used to extract natural resources, reaching a planetary scale, with more and more energy spent. As a result of this, the global-scale change in the environment witnessed in the 20th century has led to the current ecological crisis (Chapters 4–6).

As civilization has developed and especially as the global community has modernized, economic and ecological stereotypes have been formed (examples are given below) that have developed into a system of views known as "technological optimism". Though such optimism has led humankind to clash with nature and to the current ecological crisis, it still mistakenly believed by many people, expecially the ruling elite. This testifies to the urgent need for these stereotypes to be changed: they have long been obsolete and have led to destruction of both life and environment. Humankind is living under many misconceptions about itself and its relationship with nature, which have always been mistaken and do not have a scientific basis (Chapter 7).

With a new approach to calculations of the amount of anthropogenic carbon in the world, we show that humankind has overstepped the admissible limit of disturbance to the global ecosystem (biosphere) (i.e., has surpassed its ecological carrying capacity to recover). This is mainly manifested through the breakdown in stability of some environmental systems. For instance, the destruction of the elementary natural units of life (elementary units of the biological cycle) has contributed to climatic instability to much the same extent as fossil fuel burning. It is also shown that natural ecosystems, as greenhouse gas regulators, have paid a high price, which for Russia, for instance, is equivalent to the total value of all mineral resources used within its borders (Chapter 8).

In Chapter 9 it is shown that natural disasters have increased in number due to global changes and are characterized by either prolonged or long-term development.

Thus the book demonstrates the mechanisms of the functioning biosphere or "the Earth system" (we define this term as the encompassment of all life), the laws that apply to this system, and the resulting limitations of civilization development. The book shows where the neglect of these laws has led, because of insufficient understanding of the real order and functioning of the Earth system, obsolete stereotypes and myths, as well as with the much delayed initiation of studies into this most complicated system.

Any further strategy for civilization development by policy-makers at every level, for various spatial and time scales, should have a scientific basis, taking into account special features of economic, social, cultural, technological, institutional, and other aspects (Rotmans and Rothman, 2003). This book is meant for specialists in the field of Earth science and life sciences. The experience of one of the authors has shown that this book can also be used as a textbook, since a course connected with its content, though optional, has evoked great interest among the students of the Moscow State University.

ACKNOWLEDGEMENTS

We are particularly grateful to the director of the Institute of Geography, Professor V. M. Kotliakov, Professor of Moscow University, G. N. Golubev, and three anonymous reviewers for their support of this book, valuable comments, and discussions of its content. This work is supported by grants No. 00-05-64243 and No. 03-05-78009 of the Russian Foundation for Fundamental Research.

Abbreviations

DIVERSITAS	international biodiversity program
DNA	deoxyribonucleic acid
EPA	environmental protection agency
EIA	environmental impact assessment
FCCC	Framework Convention on Climate Change
GSE	Great Soviet Engelopedia
GDP	gross domestic product
GGP	gross global product
GIP	genuine indicator of progress
GNP	gross national product
IGBP	International Geosphere–Biosphere Program
IHDP	International Human Dimensions Program on Global Environmental Change
ISEW	index of sustainable economic well-being
IPCC	Intergovernmental Panel on Climate Change
NDVI	normalized differential vegetation index
SAT	surface air temperature
STI	scientific–technical information
UES	United Electrical Systems
UNCED	United Nations Conference on Environment and Development
WCRP	World Climate Research Program

Figures

Tables

Introduction

God does not play dice.

A. Einstein

Nature, be as it may,
But devil is its coauthor – that's the way

I. Goethe

Rozenberg (1999) published a paper in the journal *Ecology* in which numerous definitions of the term "ecology" were considered. He started his paper with an epigraph from the book *Ecology of Animals* by Macfedien (1965): "So what is this upstart among sciences which is only a set of facts without any theory and which obviously suffers from excess observations and lacks for principles to classify them? Does such a science as ecology exist at all?" The authors of the current book believe that by the beginning of the 21st century an answer to this question had been found and that the answer is positive. Positive because of the development of the theory of the biotic regulation of the environment. This theory did not appear suddenly. Partly it was the development of the ideas and hypotheses expressed by such scientists as J.-B. Lamarck, A. Humboldt, J.-D. Reclus, V. I. Vernadsky, V. V. Dokuchaev, A. Lotka, V. N. Sukachev, A. A. Grigoryev, V. B. Sochava, I. P. Gerasimov, N. V. Timofeev-Resovsky, and J. Lovelock, and it was partly the consequence of a huge volume of work carried out on generalization of "facts" and "observations" (as mentioned by Macfedien) or, as Vernadsky puts it, "an empirical generalization". All these experts emphasize the close connection of life (biota) with the environment, their integrity, the coupling of processes taking place in the environment, the structuring of life and habitat, and the leading role of life and matter cycles in the environment. They write about how the environment is controlled by biotic systems.

V. G. Gorshkov from St Petersburg undertook a difficult and thankless work on the empirical generalization of available information and as a result formulated the theory of the biotic regulation of the environment. We fully understand why this work is difficult. It is thankless because, like every new paradigm, it is first unaccepted and rejected. But gradually this theory has become accepted by most of the scientific community, not only in Russia but throughout the world. It is enough to quote from Steffen and Tyson's (2001) book which was the basis of the Amsterdam Declaration – the final document of the "Challenges of Changing Earth" international conference held in July 2001, dedicated to the results and accomplishment of the four largest scientific programmes: the International Geosphere–Biosphere Program (IGBP), the International Human Dimensions Program on Global Environmental Change (IHDP), the World Climate Research Program (WCRP), and the international biodiversity program DIVERSITAS. They write: "The Earth is a system that life itself helps to control. Biological processes interact strongly with physical and chemical processes to create the planetary environment, but biology plays a much stronger role in keeping Earth's environment within habitable limits than it was previously thought" (Steffen and Tyson, 2001).

Life sustains itself by making the environment optimal for itself. Therefore ecology is the science of the stability of life (biota) and stability of the environment. Life is a truly self-regulating system and controls not only chemical, but also important physical processes in the living space on Earth. Like any physical laws, the laws resulting from this theory are obligatory both for natural biota and for humankind. They put certain limits on the physical growth of civilization.

The theory of the biotic regulation of the environment is applicable to the so-called biological (classic) ecology and to the interaction between "society and nature", or simply ecology in the broad sense of the term, which Western scientists call environmental science and Russian scientists call ecology or social ecology. This theory is a uniting force, which helps us to understand the many phenomena and processes in geography, biology, and ecology, and the functions and hierarchy of the structural units of life space (the habitat for all living organisms, including humans), which are studied by biologists, geographers, and ecologists, as well as enabling us to understand the associated terms. Also, the theory represents the structural units of life space (the cells of the biochemical cycle, according to Timofeev-Resovsky, 1968) – those parts of the mechanism that are responsible for regulation and stability of the environment and, hence, determine the ecological stability of humankind, but not the laws of the human community and their technical systems.

As far back as 1944, Vernadsky wrote in his paper ("Some words about the noosphere") about the impossibility for humankind "to freely build his own history", emphasizing a necessity to adhere to the laws of the biosphere, which means, in fact, that the physical development of civilization is limited. In 1968 another outstanding Russian biologist and ecologist Timofeev-Resovsky (1968) wrote that the biosphere "forms the human environment" and controls this environment. He went on to note that "people without the biosphere or with a poorly functioning biosphere would not be able to exist on Earth" (i.e., the biosphere must be preserved). Tiuriukanov, a pupil of Timofeev-Resovsky, wrote in his reminiscences that during a lecture someone asked Timofeev-Resovsky, why there was no life on the Moon, he answered: "Because there is no regulation there" (i.e., no species governing their own environment).

In the scientific and popular scientific literature the control of ecosystems, river watersheds, natural territories, biodiversity, and other natural objects are often written about. An objective individual studying such publications will notice that it is always a question of control in the interests of satisfaction of some human need – always at the expense or destruction of a part of nature. Control over natural systems with the objective of their preservation has never been considered. Control over nature is always chaotic and always destructive for nature. The history of such control, starting from the Neolithic revolution, has led to the development of a severe global ecological crisis, in which the destruction of the natural mechanism for environmental control has led to rapidly growing global changes. It is fair to say that this control is based on the knowledge of human needs and on a lack of knowledge about what needs to be controlled. So, in this work we study the scientific foundations of life stability. Without it, sustainable development is impossible. And sustainable development must be the main goal of humankind in the 21st century.

1

Ecological–geographic space: a fresh view of old problems

1.1 TERRITORY AS A LIFE SPACE

The term "territory" usually refers to areas of land, or land and water bodies together, and is understood as determining a certain part of the Earth's surface belonging to some nation, community, or human being. Often territory is considered as the space needed to maintain the life of populations of this or that kind of organism, especially gregarious animals. Territory is also considered as a most important economic resource.

To achieve an income of 1 million roubles in the Russian chemical industry (prices as of 1988) required a territory of 6–8 ha, in the mining–chemical industry this was 20–22 ha, and in the building materials industry 3–5 ha. In the 1990s, to produce 1 t grain, 0.4 ha was needed (a global average). In 2000, to get US$14,000 of gross global product required 1 ha of land.

However, the main function of territory is to be a "life space", the habitat for all living beings, including humans. As far back as 1869, Ratzel called the Earth surface a life space, the habitat of living beings. The life space includes not only the land surface but also its subsurface, water depths and the atmosphere.

On land, the thickness of the layer where the main mass of organisms is concentrated is small: a few dozen metres (from soil subsurfaces to the tallest trees). Below the soil layer, life declines both by mass and by size. The same is observed above the vegetation cover, in the atmosphere. In the World Ocean, life reaches the deepest layers, at depths of 3.6 km, the average depth of the ocean. However, most important for life is the surface photic layer (100–200 m thick), where the photo-synthesizing organisms are concentrated – mainly phytoplankton. Here the main mass of organic matter is produced, the source of the nutrient chain for most of the organisms–heterotrophs of the ocean. The whole space where life exists is called the biosphere. It is inseparable from its biological content, with which it closely interacts. Traditionally, geography studies the medium that surrounds life and interacts with it. Ecology studies the interaction between medium and life, and biology investigates life itself.

In geography the life space is called the geographical or physico-geographical envelope (Grigoryev, 1932; Gerasimov, 1985), or geographical environment. The geographical envelope and the biosphere coincide in size, but according to Vernadsky (1967) the biosphere (including the abiotic part) is created and organized by living substances, while according to Grigoryev the geographical envelope is the carrier of the physico-geographical process and is formed through the interaction of lithosphere, atmosphere, hydrosphere, and biosphere, the latter meaning the totality of living organisms.

For a long time geography did not accept the concept of the biosphere explained by Vernadsky (1967). As Dronin (1999) writes:

> Armand (1975) expressed the "physical–geographical" view of the relationship between "living" and "non-living" components of nature, when he wrote that the notion of the "biosphere" after Vernadsky is based on an overestimation of the role of living beings in the organization of the geographical environment: "Life transforms the Earth's image in a similar way as frescoes transform the interior of a building. But it is unlikely that there is an architect who would assert that frescoes support the weight of the building." He pointed out that according to Grigoryev (1934), the biosphere is not a complex sphere including several geospheres (as proposed by Vernadsky), but a special system of bodies only consisting of living organisms that exist outside and within the atmosphere, hydrosphere, and lithosphere.

Note that Gerasimov (1985) has the same view of the biosphere. He wrote that by the biosphere of the Earth Vernadsky understood it to be the totality of all living organisms on the Earth, even though this does not correspond to Vernadsky's own views of the biosphere. Reteium and Serebrianny (1985) also noted that, on the whole, geographers do not accept the concept of Vernadsky.

The underestimation by geographers and biologists of the role of life – materialized in living organisms, the natural biota of the Earth – in the formation and organization of the environment related to a lack of knowledge about the quantitative and functional characteristics of biota and their structures despite their participation in creating the oxidizing atmosphere of the planet. Another reason was the stable paradigm of biotic adaptation to varying environmental conditions based on Darwin's theory of evolution.

According to present estimates, planetary biota total 10^{28} organisms. Most are unicellular micro-organisms. With the Earth's surface area being ~ 500 million km^2, this means that there are 2×10^7 organisms per cm^2 of the surface. With the thickness of the biosphere conditionally equal to that of the photic layer (200 m), 10^3 organisms, on average, populate each cm of this layer, or 10^3 organisms per $1\,cm^3$ of the geographical environment. This living space is populated non-uniformly. On land the population is at a maximum in the soil (*The Environment*, 1993), and in the World Ocean the maximum is in the zones of fronts where cold and warm water meet (Aizatullin et al., 1984). These estimates testify to the high saturation of the geographical environment with living organisms and, hence, to the enormous role played by life in its formation. This is especially evident in soil, the environment most saturated with living organisms: the geochemical medium of the soil, for instance, sharply differs from that of the atmosphere, just as soil solutions differ from the rainfall that feeds them.

Data on the living biomass of the Earth, taken from different estimates, arrive at totals from 1,000 to 10,000 Gt (Kondratyev and Isidorov, 2001). This means that $1\,cm^2$ of the planetary surface contains 0.001–0.01 g of living matter, while on land this figure is 1–2 orders of magnitude higher in some places. This thin film of living substance during photosynthesis produces about the same mass of matter every year. Over a period of 1,000 years it might form a layer of about 1 m, and over 1 million years a layer of more than 1 km. One gram of living matter resulting from photo-synthesis requires the transpiration of 100–1,000 g of water, which is released into the atmosphere. The living biomass is distributed over the land surface non-uniformly. Its specific value (i.e., mass per $1\,cm^2$ of land surface) is several orders of magnitude higher than in the ocean, being especially high in the forests. Taking account of leaf area index, we can assume that in the forests the specific leaf biomass is four times greater, on average, than for other areas of land. Therefore, the transpiration of water in forests is much greater, meaning that land vegetation is an important regulator of the continental hydrological cycle.

Yet more important is the role of marine biota, which according to present-day estimates transport the whole volume of the World Ocean through its many organisms twice a year. Assuming that half of all living organisms (i.e., 0.5×10^{28}) dwell in the ocean, one living organism will transport through itself about 0.1 g of seawater annually. From seawater and the suspensions in it, the biota extract gases, suspensions, and solutions. This affects the concentrations of substances important for life in the oceans. In ocean water a strict ratio between basic nutrients is observed: $O:C:N:P$, known as the Redfield ratio. The same ratio is observed in the synthesis of organic matter; the Redfield ratio is not accidental, it is maintained by the activity of all marine biota.

These data show that the biological processes set in motion by biota are neither outside the atmosphere, hydrosphere, and lithosphere, nor near them, and are not "frescoes on the wall" as Armand (1975) believed. Biota dynamics is a mechanism that creates both the "wall" and the most beautiful "frescoes" on it. The work of biota on the organization and regulation of the environment is the united physico-geographical process that Grigoryev (1934) discussed. When this idea was first mentioned, there was insufficient data to confirm that individual processes in the geographical environment are parts of the united process. On the basis of intuition, Grigoryev suggested that there is a real mechanism that builds and regulates the environment. Of equal importance, he considered landscapes to be an external expression of a united physico-geographical process, though he defined the phyto-ecologo-geographical process as but one of its components. Grigoryev (1934) believed that landscapes might be selected as an indicator of a physico-geographical process, quantitatively, by the estimates of the vegetation produced, dead material (referring to pure primary production), etc. In his paper "On chemical geography"

(1936) Grigoryev wrote: "Of importance is not the elementary chemical composition of the landscape, but revealing the processes that create the chemical structure of the physico-geographical environment." In 1943, putting forward the law of intensity of physico-geographical processes, Grigoryev attached greater importance to biotic activity, noting that the intensity of this process is expressed through a cycle of living organic matter and through a cycle of mineral substances. Sochava (1970), in search of factors contributing to the development of the physico-geographical process, wrote about the integration and interaction of natural regimes resulting in a kind of landscape-forming effect. It is clear that the integrating element is the biota, functioning as a regulator and stabilizer of the environment.

Gerasimov (1985) believed that one of the main problems of geoecological monitoring was geosystem monitoring based on studies of the internal cycle of matter in the main natural ecosystems, as well as its anthropogenic transformation. In this work, Gerasimov dedicated a special chapter to problems of the internal cycle of a substance in the ecosystems for the Russian territories. He pointed out, in particular, that a sound typology for natural ecosystems had still to be developed. Therefore, it is necessary to understand the essence of basic processes and, in particular, the natural cycles of substances taking place in various natural eco-systems. Thus the top Russian geographers considered the structural units of life space, biotic produce, and the biogeochemical cycles of matter, when the problem arose of making quantitative assessments of the processes taking place in geosystems and of geosystem dynamics.

Gerasimov (1985) also pointed out that due to deforestation the intensity of natural cycles of matter (nutrients) breaks down. This leads to the weakening of biogenic cycles that are necessary for the natural annual level of primary and secondary biological productivity, resulting in their decrease and, hence, a break in their natural balance and their rate of formation. As for agriculture, Gerasimov notes that crop growing destroys the surface biomass of a natural ecosystem, with subsequent loss of most of the primary annual biological production. All these changes result in breaking both the natural balance of biological production and its rate of formation (restoration). Depending on the degree of damage to above-ground residual biological production, the size of lost annual primary production, and the artificial increase of nutrients, the effect on the ecological balance in the natural ecosystem can vary. As for pasture stockbreeding, Gerasimov also noted that it leads to disruption of the ecological balance and the digression of pastures.

Finally, we should mention the work of Sochava (1978) in which he singled out biota as one of the main components of geosystems, along with solar radiation and water. However, it is now clear that land biota control 70% of the continental water cycle and affect the reflectivity (albedo) of the planet (i.e., to some extent, they govern the physical processes of the geographical environment).

So, Russian geographers have attached great importance to the cycle of nutrients in ecosystems and the balanced nature of this cycle. They understood its leading role in ecosystem dynamics, and from geosystem monitoring showed that human activity disrupts the balance in geosystems, leading to their instability. As a result of the destruction of natural ecosystems the cycle of nutrients breaks down,

and to restore it (e.g., on agricultural lands) considerable additional energy investments are needed.

Armed with this knowledge, it seems amazing that the geographers and biologists of the 20th century, who thoroughly understood the role of humankind (also
part of living substance) in the reorganization of the world and creation of artificial
agrosystems and technosystems, did not attach enough importance to natural biota
in the rearrangement and reorganization of the biosphere or geographical envelope.
The cycles of nutrients and the functioning of biota constitute the main mechanism
that creates and maintains the natural environment of humans and all other living
creatures (Gorshkov, 1995; Gorshkov et al., 2000).

With the advent of the Neolithic (agricultural) revolution, the roots of the
modern market economy appeared, in which humans organized new, unbalanced
cycles of various substances (nutrients included), increasingly violating the natural
biogeochemical cycle. In the 20th century this anthropogenic cycle, first local and
then regional, became global in scale. At the same time, in the process of human
activity, the balance of natural biogeochemical cycles was violated, destroying or at
least strongly deforming natural ecosystems. Grigoryev (1934) was right in introducing the notion of a complex geographical process that includes economic, social, and
physico-geographical components. Now we can see the results of its action in the
form of a deepening social–economic–ecological crisis.

The concept suggested by Grigoryev (1934) of a united process was criticized
toward the end of the 1940s, and in the 1950s the critique was in full swing. There
were several reasons for this. First, the author was unable to define the notion of a
united physico-geographic process for lack of sufficient scientific data. Second, in all
geographic discussions at this time a lot of communist ideological pressure prevailed.
All sides used this as a means of relegating the scientific approach to the background.
Third, there was the erroneous statement that landscapes are only an expression of a
united process, while in fact they are the basic mechanism of the process. Fourth, the
division of geography into numerous individual disciplines played did not help. And
fifth, there was a significant gulf between biology and geography.

We agree with Isachenko and Solntsev (epilogue to the book by Dronin, 1999)
that geography is created not in academic circles but in the process of practical
studies, and that:

> each moment and stage of the progress of geographical thought appearing in the
> form of "university" or "academic" publications ... is only an external expres-
> sion of deep processes of irreversible accumulation and generalization of initial
> geographical information.

So, both the geographic envelope and the physico-geographic process as proposed by
Grigoryev, and constructive geography as proposed by Gerasimov have not been
determined by ideological or subjective considerations. In the light of newly gathered
data, the ideas mentioned above should be reviewed and definitions of the terms
transformed. Geography and biology must come gradually to an understanding of
the unity of the processes taking place in life space, which life itself ensures. In this

process, life unites all media and all natural organisms. It is life itself that ensures this process, by structuring its space.

1.2   GEOGRAPHIC ENVIRONMENT, GEOGRAPHIC ENVELOPE, AND BIOSPHERE

The term "geographic environment" has long been used. It was understood to mean the natural conditions in which humans and all human communities live and act. Barkov (1954) writes: "physical geography studies the natural conditions surrounding human society – the geographical environment that consists of the Earth's crust, troposphere (the lower atmosphere), water, soil cover, vegetation and animals." In the *Great Soviet Encyclopedia* (*GSE*, 2nd edn, vol. 10, 1952) the term geographic environment was defined as "the nature surrounding society is one of the needed and constant conditions for the material life of this society." Later on, in his paper "Geography" (Grigoryev in the same volume of *GSE*), the definition of the term geographic environment coincides with that by Barkov.

In 1956, in *GSE* (2nd edn, vol. 45, 1956), Grigoryev in his paper "Physical geography" uses the term "geographical envelope" as a broader notion, mentioning that the French geographer Reclus (1876) defined the geographical environment as "interconnected natural phenomena surrounding the human society." In *GSE* (3rd edn, vol. 6, 1970) the notion of geographic environment is narrowed from the nature surrounding society to "a part of the terrestrial natural ... environment of human society, with which society is directly connected at a given moment in its life and activity." The same definition of the term geographic environment is used in *Protection of Landscapes Glossary* (1982), but it is suggested that the term be qualified according to which environment is meant – natural or anthropogenic – and to which subject it refers (society, settlement, etc.).

In the reference book *Nature use* Reimers (1990), gives two definitions of the term geographic environment. The first definition repeats the definitions of the last edition of *GSE* (vol. 6), and the second definition takes into account the comment in *Protection of Landscapes* (1982): "a totality of natural, technical, technogenic, social, and economic conditions of life." At the same time, Reimers notes that the geographical environment is very close to the "natural environment at a large scale."

In one glossary – *Geoecological Russian–English Dictionary* (Timashev, 1999) – we return, in fact, to a broadened interpretation of the term geographic environment – "the Earth's nature in which humankind lives and develops is either virginal or changed to some extent due to man's activity." This latter definition again broadens the notion of geographic environment. But, for reasons unknown, at the end of definition the author notes that the geographical environment is part of the geographical envelope. The broadening of the content of the term geographical environment is quite natural, since, in fact, humankind always lived and developed in the geographical environment after the definition by Barkov (1954), and the interaction of humankind with the geographical environment included lithosphere, soils, troposphere, natural waters, vegetation, and animals, both on planetary and local scales. For instance, such phenomena as earthquakes, whose centres are at depths down to the base of the lithosphere, always affect the development of humankind. Now man is actively interacting with extraterrestrial space.

What is the relationship between the notions of geographic environment and geographic envelope? The latter term appeared much later, in the 1930s, and was suggested by Grigoryev (1932). As the author of the term notes, it developed from the ideas of Dokuchayev (1948) who believed that the Earth's crust relief, climate, waters, vegetation, animals, and soil cover are closely connected and correlate, forming a single whole to be thoroughly studied. Grigoryev proved that the outer mantle of the Earth has qualities of its own and therefore should be studied as a special phenomenon of nature formed on the Earth's surface, and called this space the physico-geographic envelope. In fact, he equalized the notions geographic environment and geographic envelope, noting in his paper "Physical geography" (*GSE*, 2nd edn, vol. 45, 1956) that "this envelope is the habitat of humankind and the place of its activity", which Reclus (1876) called geographical environment. Grigoryev, instead of transforming the definition of the term geographical environment, which is quite natural for developing science, introduced a new term, which was geographical envelope.

Reimers (1990) gives the following definition of the term geographical envelope – "a natural complex appearing in the layer of interaction and mutual penetration of lithosphere, hydrosphere, and atmosphere and formed under the influence of solar energy and organic life. As a rule, the geographical envelope includes a 10–12-km layer of the atmosphere, the whole hydrosphere, and a 4–5-km layer of lithosphere." A definition by Timashev (1999) is close to that by Reimers, only the role of the internal energy of the Earth is added, but the role of life as a forming force is omitted. These interpretations determine not only the boundaries of the geographical envelope but also its specific features: the presence of life in it, the possibility of the existence of water in three phases, territorial differentiation, and formation and existence due to solar energy and life functioning.

Thus the geographical envelope by spatial coverage (lithosphere, hydrosphere, soils, atmosphere, animals, and vegetation) is similar to the geographical environment. In its definition, conditions of its formation and functioning (solar energy, geothermal energy, and life) are additionally introduced, and it is pointed out that this emphasizes the features inherent to the geographic environment. Everything written above leads to the conclusion that the terms geographical environment and physico-geographic envelope or geographic envelope are synonyms. It is clear, why the term geographic envelope ceases to be used, and geographic environment remains.

Grigoryev (1948a) emphasized the role of solar radiation in the formation of the physical medium of the geographical envelope (geographical environment) and pointed out that life exists only in this physical medium, but he did not take into account the role of life in the organization and formation of the geographical environment (envelope). Only in the definition of the geographical envelope (environment) by Reimers (1990) is the role of life in its formation noted.

As noted in the work *Protection of Landscapes* (1982), instead of geographical envelope other terms were proposed: "epigeosphere", "geographical sphere", and "biogeosphere". But they have not become widespread. Besides, epigeosphere (epi – above something) narrowed the term geographical envelope (environment). The term geographical sphere changed nothing, except the word "envelope" for "sphere". The term biogeosphere combined geographical envelope with the biosphere, but the latter, as was mentioned in Section 1.1, during the period of formation of these terms was poorly accepted by the geographic community.

The term "biosphere" was proposed by Suess (1875), when he developed an idea of concentric envelopes of the globe. Vernadsky (1967) defined the biosphere as the Earth's envelope occupied by a totality of living organisms (living substance), including the troposphere, hydrosphere, and upper lithosphere. Though the studies of Vernadsky on the biosphere appeared as early as the 1920s, the physico-geographic dictionary by Barkov (1954) gives a definition of the biosphere suggested by Suess as "a totality of organisms populating the Earth". In the book *Protection of Landscapes* (1982) the authors noted that the geographers' interpretation of the term was "spatial-substrate" (after Suess, 1875). The biosphere was considered one of the geospheres forming the geographical envelope along with the atmosphere, litho-sphere, and hydrosphere, but differing from them by the extent of saturation with living organisms (vegetation cover, animals, sphere of biocenoses) and with products of their vital functions. Further it was mentioned that after Vernadsky, the biosphere began to denote the whole outer sphere of the planet, in which not only life exists but which is somehow changed or formed by life, and then "by their substrate character-istics, indications to the presence of substance in three aggregate states, by the list of components, taking part in its construction, the notion biosphere is close to the notion geographical envelope". In the compilers' opinion, the difference is that in this term a special role of life is emphasized and put forward.

Vernadsky (1967) wrote that the biosphere "is confined, first of all, to the field of life existence." This field of life existence includes the troposphere, hydrosphere, and lithosphere down to several kilometres, probably, to the boundary of transformation of water to plasma at the temperature of $460°C$. At the boundaries of the field, life is strongly rarefied. The highest density of life is observed in soils, on land surface, in the upper photic layer, and in the regions of fronts of the World Ocean.

Thus in definitions of all three terms mentioned the subject of studies, or their object, is the lithosphere (its upper part) together with soil, hydrosphere, lower atmosphere, vegetation, and animals. Naturally, the boundary of this object of study can broaden with the development of science. From the time when Reclus (1876) introduced the term geographic environment (at the same time, Dokuchaev (1948) pointed out the genetic integrity of the natural environment), the definition of this term has been constantly supplemented. Grigoryev (1965) showed that the integrity in the geographical environment is determined by assimilation of solar energy: "processes in it take place due to both cosmic and terrestrial sources of energy (in some spheres due to one of them)"; and "there is life in it." The latter emphasizes one of the most important features of the geographic environment. Another feature – the geographic environment is structured. Finally, he points to one more feature – the presence of water in three aggregate states. Thus the problem concerns specification of the object of study. Apart from material objects, processes and features of this object are studied, which ensure its genetic integrity. Therefore, there was no need to introduce a new term geographic envelope. Reimers (1990) specified this notion (also the geographic environment). He emphasized that it was formed under the influence of solar energy and organic life.

Now we can formulate a new definition of the term geographical environment:

A special sphere of the Earth where the biota-formed upper part of the lithosphere, hydrosphere, soil, and atmosphere (troposphere) interact and where the processes take place due to biota, solar radiation, and, partially, internal forces of the planet. Its special features are the presence of life, water in different phases, structuring of life, and space occupied by life. Within this sphere, and due to it, there exists and develops humankind, which has begun to markedly influence the formation of the geographical environment and processes taking place in it.

Bearing in mind the role of life (biota) in the formation of the geographic environment and of the processes taking place in it, this definition can be simplified as follows: *the biota and the environment surrounding it.*

Terms and their definitions in science

Natural sciences are based on experiments, therefore their basic terms are chosen so that they permit continuation of scientific studies. In mathematics this is impossible, because terms are based on the formulated postulates, and therefore are strictly defined. In physics, for instance, the definitions of terms can be changed. Sometimes, instead of transformation of the definition, scientists suggest a new term with a new definition. Such was the case with the term geographic envelope introduced instead of geographic environment, though it would have been sufficient to transform the definition of this term. In natural sciences, studying nature and the phenomena taking place in it, definitions of the terms can and should be supplemented and changed with an accumulation of knowledge.

Coming back to the term biosphere, one should note that it appeared as a purely biological notion – a totality of organisms. Vernadsky (1967) developed the concept of the biosphere as a life space and its geochemical role. Thus, this term was included in the system of geological sciences, and now it is one of the basic terms in ecology. The dictionary of geographic terms by Barkov (1954) lacked this term, but now it is included in all geographic and ecological dictionaries and encyclopaedias. Thus, the content of the term biosphere was constantly changing. The definition of this term (*Protection of Landscape*, 1982; Timashev, 1999) is close to that of the geographical envelope and hence, geographic environment. In *Protection of Landscapes* (1982) interpretations of biosphere and geographic environment (envelope) differ not only by an indication of high concentrations of living organisms, but also they emphasize a special role of life. Therefore, in the definition of the geographical environment it is necessary to take into account the role of life as a most important mechanism for the formation of this environment and processes taking place in it, not only biogeo-chemical, but also physical (e.g., continental water cycle, weathering, alluviums transport, changes of surface albedo, etc.). Thus, the present definition of the biosphere is similar to that of the geographical environment given above. To put it briefly, the biosphere incorporates the biota and its environment.

1.3   ENVIRONMENT

The term "environment" is interpreted equivocally in Russian literature. It appeared in Russia in the 1960s, but now it is widely used, including in official documents of the Russian Federation. Naturally, it is absent in the dictionary by Barkov (1954). In *Protection of Landscapes* (1982) this term is interpreted as human's surroundings, with the accompanying comment that it is an anthropocentric notion: "it is used to show that the notion concerns itself with life conditions, the habitat of a human being or (in the territorial sense) population …. Clearly, not only natural phenomena form the domain surrounding a human."

Reimers (1990) writes that the environment is the same as the external surround-ings but is in direct contact with an object or a subject. He believes that the term should be more specific, including: what or who is surrounded. Therefore, he suggests two terms, believing that the term "environment" without additions does not correspond to the semantics of the Russian language. The first term "human's surroundings" includes not only abiotic and biotic media, but also a social medium, together and directly affecting both humans and the economy. The second term is "human's natural surroundings". This term has three definitions. The first definition includes a totality of natural as well as abiotic and biotic natural factors changed slightly by human activity and affecting human beings. In subsequent definitions, the agricultural territories transformed by man's activity are added to a totality of natural factors. In these broadened definitions the notion of human's surroundings merges with the notion of human's natural surroundings.

Timashev (1999) defines the last two terms more clearly. The human's surround-ings include a totality of natural abiotic and biotic conditions, which affect human beings and other organisms. The human's surroundings are a totality of natural (abiotic and biotic), natural–anthropogenic, technogenic, and social conditions in which humankind lives and will continue to live. Then it is added that the environ-ment is a part of the geographic environment, but it is not explained, why it is only a part nor does it specify which part.

The definitions of the notion of environment given above do not indicate why their authors keep trying to place a human being at the centre of the definition. In the English language the meaning of the word environment is clearly defined. *The Oxford Desk Dictionary and Thesaurus* (1997) gives the following definition:

"surroundings, especially affecting life and conditions, or life surroundings", and then follows a list of terms including the words atmosphere, territory, habitat, biosphere, and ecosystem. This list does not contain any socioeconomic terms. Hence, the environment is a totality of abiotic and biotic natural conditions affecting life (not only of human beings). Apparently, man's activity can change these natural conditions, but they still remain the environment. As for economic and social conditions, they should not be included in the definition of the environment. It is also clear that human beings are also included in the notion of life, and the environment refers, on the whole, to life or to its component parts.

In the dictionary by Porteous (1996) the term environment is defined as "all the surroundings of an organism, including other forms of life, climate, soil, etc. In other words, conditions for development and growth."

The term environment in English is now widely used in the Russian scientific literature, therefore the discrepancy between Russian and English semantics, as Reimers (1990) wrote, is now absent. Proceeding from the original sources, this term should be understood as natural life surroundings, including those changed by human activity. Life can be considered as a whole, as individual communities of organisms, and as simply organisms, including human beings and human communities. The surroundings (*The Oxford Desk Dictionary and Thesaurus*, 1997; Porteous, 1996) include the lithosphere together with soil, as well as hydrosphere, atmosphere, and other organisms, if life is not considered as a whole. This definition brings together the terms biosphere, geographical environment, and environment, which function as synonyms being used by specialists in different fields: geographers, biologists, and ecologists (Table 1.1). This list does not contain the term geographic envelope.

The process of formation of such terms/synonyms in various fields of science is natural, especially in the sphere of Earth and life sciences, since all life sciences study, in fact, the same region or sphere of our planet, as do the Earth sciences. Division of sciences is a kind of conventionality appearing for convenience of studies and education of specialists, since, first, one cannot seize something boundless, and, second, deeper studies are needed of narrow spheres in some disciplines, especially for practical purposes. In the late 20th century a necessity arose for broad generalizations. Ecology, appearing at the turn of biology and geography, is one such generalizing sphere, and within geography itself appears landscape science. Because of this diversification of subjects it is necessary to

Table 1.1. Comparison of the terms geographic environment, biosphere, and environment.

Term	Forming objects	Basic processes
Geographic environment	Lithosphere, soil, hydrosphere, atmosphere, various forms of life	Solar radiation, functioning of living organisms, internal forces of the Earth
Biosphere	Same	Same
Environment	Same	Same

compare terms and definitions in order that specialists in various fields of science
could understand each other.

1.4   FORMATION OF THE GEOGRAPHIC ENVIRONMENT

The present ideas of the role of life in the formation of the geographic environment
proceed from the studies of the naturalists of the past. "Lavoisier and Sniadetsky
were first to bring forth a hypothesis of the existence of biogenic cycles of chemical
elements, while Lamarck established in the early 19th century that all complex
organic substances observed in nature are present in living organisms. Therefore,
he reached the conclusion that: all minerals in the outer crust of the Earth are
exclusively products of the vital functions of animals and plants, which had
existed at some time on the Earth's surface. In continuation of these ideas,
Humboldt (1826) defined the organic world as an integral part of the Earth's crust
and pointed out the dependence of the chemical composition of the organisms on the
inorganic matter of their habitat. He introduced the notion of 'vital environment' as
a specific envelope of the Earth, which combines the atmospheric, marine,
continental processes, and phenomena connected with the vital functions of the
organisms" (Ivanov, 1999).

There is enough convincing evidence of the formation of the geographic environment (the biosphere and the environment) by life (biota). Geology admits that most
of the Earth's lithosphere is formed through direct or indirect participation of biota.
In particular, life has favoured the formation of minerals such as the largest deposits
of iron, numerous deposits of phosphur, and bauxite. Biota has promoted the
formation of carbonates constituting almost 20% of the sedimentary formation of
the lithosphere. The granite-metamorphic part of the lithosphere, which Vernadsky
(1967) called "footprints of the ancient biosphere", is now known to have been
created with the participation of life (Lapo, 1987).

With an appearance of life on land, the process of weathering accelerated by a
factor of several hundred, since biota makes the basic contribution to this process.
Weathering is, first of all, a biochemical process. Experimental data show that
vascular plants, for instance, accelerate weathering by a factor of 10–100 (in this
case not only detritus appear but also solutions), and moss and lichen accelerate
weathering by a factor of 100–300 (Lapo, 1987).

The appearance on land of the continental water cycle is connected with the
transport of life from the oceanic waters to continents. Life, which originated in the
ocean, colonized the land. It created the soil cover, which is a porous medium
holding moisture. The total depth of this "ocean" constitutes about 10 cm of
water. Life also created on land an enormous evaporating surface in the form of
leaves, stems, and needles (the area of which is no less than the area of the World
Ocean). The area of the Earth's surface covered with vegetation evaporates moisture
mainly via the process of transpiration and "catching of precipitation" by the plants'
canopies. Controlling 70% of the continental moisture cycle on land, living
organisms thereby take part in sinking the deposits and dissolved substances into

the ocean, providing the geological cycle of substances. The tropical forests create a very powerful local moisture cycle, favouring a washing of the weathering crust bringing substances into the ocean.

Biota may have played an important role in the formation of the water masses (oceans). At present, concentrations of basic nutrients in the oceanic waters are maintained by marine biota, which twice a year transmit the volume of the oceanic water masses through the bodies of its constituting organisms. As a result, molar ratios of dissolved inorganic carbon, nitrogen, phosphorus, and oxygen in the ocean coincide with the stochiometric ratios of these elements observed in biochemical reactions of synthesis and decomposition of the organic material of marine biota. This ratio for C, N, P, and O constitutes 106/16/1/138 (Redfield ratio).

The present oxygen atmosphere has been created with direct participation of biota together with the ozone screen, which protects life from harmful UV rays.

The Earth's soil cover (completely created by biota), is a component with its specific gaseous and aqueous medium. The air within the soil cover is closer (by composition) to ancient atmospheres. Biota forms soil solutions specific to different types of soils. In $\frac{1}{3}$ m^3 of soil there are up to 1 trillion micro-organisms, and in 1 cm^3 up to 2 km of fungus hyphae of micron thickness. Such relatively large soil organisms as rain worms, over a period of 100 years transmit a soil layer 0.5 m thick through their stomachs (Lapo, 1987; *The Environment*, 1993).

These and other examples illustrate how life, materialized through the sum of natural organisms – biota, controls the fluxes of substances using solar energy, and builds and organizes an environment of its own. The magnitude of energy used by biota is given in Table 1.2, which shows that solar energy generates the basic physical processes on the Earth's surface: circulation of the atmosphere and ocean as well as the global water cycle. Solar energy is also a source of energy for the main part of the process of evaporation from the land surfaces. The share of biota in the total evaporation (evapotranspiration) from the land surface constitutes about three-fifths and, hence, its contribution to precipitation and transport of alluviums and dissolved substances from the continent to the World Ocean is of the same order.

Based on the above, one can state that there exists a united process in the geographic environment (biosphere), which is provided by two factors: (1) by the supply of solar energy for the physical processes in the geographic environment; and (2) by biotic activity, which, using solar energy, is responsible for biogeochemical processes on land and in the ocean as well as participating in various physical processes.

These processes result in the formation and regulation of the geographic environment of the planet (i.e., biosphere, geographic envelope, and environment). Over billions of years of the existence of life this united mechanism, while continually transforming and changing, maintained conditions favourable for life. Therefore, the notion of a united physico-geographic process introduced to the field of geography by Grigoryev (1932) cannot be considered a tribute to the "formal use in geography of the general principle of dialectic materialism of the forms of matter motion" (Dronin, 1999). At that time there was no sufficient information for deep scientific generalization, therefore the role of the biota in the formation of its own

Table 1.2. Energy fluxes onto the Earth's surface (Gorshkov et al., 2000).

Power sources and sinks	Power (10^{12} W)	Ratio to solar power
Solar dissipation and its power:		
Total flux from the Sun onto the Earth	1.7×10^5	1
Power reaching the Earth's surface	8×10^4	0.47
Evaporation from the Earth's surface	4×10^4	0.24
Evaporation from land surface (evapotranspiration)	5×10^3	3×10^{-2}
Poleward heat flux from the equator:		
Atmosphere	3×10^3	2×10^{-2}
Ocean	2×10^3	10^{-2}
Wind power:	2×10^3	10^{-2}
Energy of oceanic waves	10^3	6×10^{-3}
Maximum hydraulic power of rivers	3	6×10^{-5}
Power of winds and rivers available to mankind	1	6×10^{-6}
Moon light	0.5	3×10^{-6}
Transpiration by biota	3×10^3	2×10^{-2}
Photosynthesis	10^2	6×10^{-4}
Non-solar power sources:		
Total flux of geothermal heat	30	2×10^4
Volcanoes and geysers	0.3	2×10^{-6}
Chemosynthesis of life	10^{-4}	6×10^{-10}
Power consumption by man, late 20th century:		
Power consumption (fossil fuel)	10	6×10^{-5}
Consumption of pure primary production of biosphere	9	6×10^{-5}

surroundings had not been assessed despite certain indications of such a situation existing in the studies by Vernadsky (1967). The climatic constituent was considered to play the leading role in the united physico-geographic process. The role of natural biota is still underestimated in geography.

Nowadays it is more correct to speak not about a united physico-geographic process but about a united ecologo-geographic process, also considering biotic and non-biotic constituents of the geographic environment and believing that the biotic component is the leading and forming one in this environment. This process cannot be chaotic, it is accomplished through natural spatial structures that combine living and non-living elements based on physical laws and, first of all, on the law of conservation, as well as on biological laws, primarily, the law of competitive inter-action. Geography and biology revealed these structures long ago and now are studying them.

2

Geographical environment: structure and processes

2.1 ABIOTIC COMPONENTS OF THE GEOGRAPHIC ENVIRONMENT

Historically, geography has studied the space of human life. In particular, physical geography was interpreted as the science of physical phenomena in the geographical environment. Present-day physical geography cannot further allow an underestimation of biota's role in environmental dynamics. It would probably be reasonable to introduce in this context the term "ecologo-geographic environment", but the authors do not intend to increase the number of notions and will further use the term "geographic environment" and, as its synonyms, the terms "geographic envelope", "biosphere", and "environment".

The geographic environment is an open physical system in which various processes take place. As a result of energy consumption from outside the system, so-called physical self-organization takes place in this system. However, the organized processes such as physical self-organization do not take place in every open physical system but only in those that receive an information flux from the outside together with energy.

Below we shall mainly follow the studies by Gorshkov et al. (1999, 2002). Information needs memory cells. In a written text these cells are represented by the locations of letters and spaces. The letter information can be recoded into digital format, which is used in present-day computers. In computers the memory cells are macroscopic elements with various physical nature. In nature, natural memory cells are molecules of environmental matter, with which solar radiation interacts. Solar radiation transforms molecules of a substance into an excited state with the energy, exceeding the thermal (chaotic) energy of molecules. Disintegration of such an excited state leads to the generation of all physical processes observed in the environment – water cycle, wind, atmospheric and oceanic circulation, evaporation, precipitation, river flows, cyclones, hurricanes, etc.

An excitation of environmental molecules by solar radiation above the thermal noise means that the Sun sends to the Earth not just heat but an information-rich energy flux. An amount of solar energy received by the Earth is equal to an amount of thermal energy emitted by the Earth into space. However, these two fluxes differ substantially in their information characteristics. Solar radiation consists of discrete particles – photons, whose average energy is in proportion to the absolute temperature of the radiation source. It is 6,000 K for the Sun and near 300 K for the Earth (less than the Sun by a factor of 20). This means that each solar photon contains 20 times more energy than a thermal photon of the Earth. Since the fluxes of absorption and emission for the Earth are equal (otherwise, it would either cool down or warm up), each solar photon disintegrates into 20 thermal photons. In the process of disintegration, the transfer of information to memory cells takes place (i.e., generation of all the observed processes mentioned above).

If one solar photon can excite up to 20 molecules, with each molecule emitting one thermal photon, then the number of molecular memory cells should be equal (by an order of magnitude) to the number of thermal photons emitted by the Earth into space. Their number S is determined by solar radiation absorbed by the Earth's surface, $Q = 10^{17}$ W divided by the average energy of a thermal photon emitted by one molecule. This energy is equal to $K_B T$ 10^{-21} Joule molecule^{-1} (where K_B is the Boltzmann constant) and is proportional to the inverse value of the Avogadro number (6×10^{23}) ($T = 300$ K is the absolute temperature of the Earth's surface). We obtain $S = Q/(K_B T) = 10^{38}$ molecules s^{-1}. If we assume that approximately one state of a molecule is excited above thermal noise (i.e., a molecular memory cell contains two states), then the "molecule" can be substituted for a "bit". Thus, the information flux received by the Earth from the Sun has an order of magnitude 10^{38} bit s^{-1}.

Not all energy can lead to excitation of molecules above the thermal noise. If the Sun's temperature dropped down to the Earth's temperature and sent a constant flux of heat, then the Earth's surface would always remain warm, but the information flux would be equal to zero, and all the regulated processes would stop. The Earth would be in thermal equilibrium with the Sun and become its subsystem.

The energy of all regulated processes on the Earth dissipates and transforms into heat (thermal photons). Hence, information in molecular memory cells with erased information determines an extent of chaos called entropy in physics. And the cells with preserved memory, together with those without memory, are the information capacity of the system (i.e., the latter is the sum of the amounts of information and entropy). On the whole, if the Sun sends to the Earth a pure flux of information, the Earth emits a pure flux of entropy into space. Proceeding from the above, one can state that during the geological time period the processes of physical self-organization in the geographical environment could have accelerated with an increasing intensity of radiation, since from astronomical data the Sun's luminosity has grown by 30% over this time.

Transformation of solar energy into thermal energy on the Earth has always been spatially non-uniform, causing gradients of temperature and pressure, as well as irregular cycles of water. If some processes take place due to the direct impact of solar radiation (e.g., evaporation), then processes such as hurricanes, vortices, and avalanches are the result of an accumulation of potential energy (latent heat or gravitation energy). The energy is accumulated until a certain critical level is reached. This process is called a "self-organized critical state". Information about these processes is recorded not in molecular but in macroscopic memory cells, often called "degrees of freedom". For instance, in a river flow the degree of freedom of the turbulent current is determined by the amount of vortices of different sizes, and energies, which are transformed into vortices of a minimum size, and eventually into heat. The number of macroscopic memory cells is (a large number of orders of magnitude) less than the number of molecular memory cells existing in the environment. On the whole, the efficiency of transformation of solar energy into the energy of macroscopic processes of the physical self-organization in the geographical environment does not exceed 1% (Table 1.2).

Specific features, and the course of all processes of physical self-organization, are totally determined by external energy fluxes coming to the system (for the Earth – solar energy) and by the environment in which these processes take place. Cessation or change of fluxes either stop or change the processes of physical self-organization, and with the renewed energy flux they appear again (i.e., with a constant energy flux there is no evolution in the processes of self-organization). Thus, all these processes are not self-organized but organized by quantities and characteristics of the external energy flux, as well as characteristics of the environment. Hence, it is clear that the environment under the influence of energy fluxes determines the processes of physical self-organization and also controls them. Hence, these processes themselves can neither directly change the environment nor control it.

2.2 BIOTIC COMPONENT OF THE GEOGRAPHICAL ENVIRONMENT

Life materialized in biota differs sharply from abiotic objects: there are no transition forms between them. They are separated by a gap of qualitative and quantitative differences, though a detailed analysis of living organisms has shown that all processes in them follow the laws of the abiotic medium. Therefore, it is difficult to find a unique definition separating living from non-living objects.

Such differences as the requirement for food, ability to reproduce or self-preserve, and the processes of thought are not observed in abiotic media, but not all living organisms have them. Nevertheless, the differences of living objects from non-living ones are well known (Gorshkov et al., 2000).

First, it is well known that living objects are exclusively complicated compared to the self-organizing most complicated objects in abiotic media (hurricanes, tornadoes, ball lightning, etc.). Most complicated abiotic objects appear spontaneously under the influence of an external energy flux. Living objects, in the presence of the same energy flux, never happen spontaneously. Pasteur was the first to establish this fact (Pasteur's law). Now, when a huge data set has been accumulated about the complexity of molecular structures of living objects, it is clear that Pasteur's law is exclusively important in distinguishing between living and non-living objects. A further confirmation of this law is the observed fact that species that disappear (become extinct) never appear again.

Second, an important feature of living objects is their existence in the form of organisms with strictly determined sizes of bodies, inside which all structures are closely correlated (with no correlation between organisms themselves). The death of one organism does not affect either the organisms of the same species or other species, but violations of the internal structure of an organism may lead to its change and death. Ordered processes in non-living nature, like wind, sea currents, etc., can be of different sizes and spatial extents. Local violations of such processes do not lead to their destruction in other local areas.

Third, a distinctive feature of life is an organization of living objects into populations. There are no biological species consisting of a single organism. A population consists of interacting but not correlated species, and this means that the death of one or several of them does not affect the population on the whole. This feature of life differs from highly-ordered physical objects. For instance, a ball lightning, being spontaneous, disintegrates before another ball lightning appears.

Fourth, a common feature of life is its trend to expand, when each biological species propagates over suitable territory. Expansion is observed in all biological species, including viruses and humans.

To summarize, the basic distinguishing features between living and non-living objects are (Gorshkov et al., 2000):

- *A very high level of complexity of living objects compared to non-living ones and impossibility of their spontaneous appearance with any fluxes of external energy (Pasteur's law).*
- *The existence of life in the form of living organisms with clearly determined body sizes, inside which all structures are strictly correlated. Between organisms there appears no correlation (the death of one organism does not affect either its own or other species).*
- *The existence of living organisms is only in the form of populations: there are no species consisting of one organism.*
- *The trend to expand is inherent to biological species (covering the whole territory suitable for them).*

Living organisms and their communities accumulate information in molecular memory cells of a genome – deoxyribonucleic acid (DNA) – a polymer double-strand macromolecule having no limiting threshold of accumulation (genome is a totality of all genes of an organism with records of all its properties). In the process of evolution they have assimilated such a volume of information that the highly ordered processes of life determined by this information can neither be generated spontaneously nor maintained by any fluxes of external energy (Pasteur's law). *Therefore only evolution of living organisms facilitates their true self-organization.*

The preservation of genetic information and maintenance of stability of the species of organisms and their communities is based on a new principle lacking in the systems of physical self-organization. This principle consists of the formation of a population of organisms of one species, inclusion of a competitive interaction between organisms, as well as selection and exclusion from the population of individuals with partially erased genetic information (decay individuals). The ability of organisms to reproduce makes it possible to fill in the vacancies and to provide stability to the size of the population.

The level of organization of living systems is independent of the external fluxes of energy feeding them. Therefore, life has the possibility of controlling environmental conditions (i.e., changing these conditions and keeping them stable and optimal for life). The consumption of external energy – food – is only necessary to replenish energy expenditures for competitive interaction, reproduction, and environmental control.

Apparently, *a separation of living nature from non-living nature had occurred at the moment when life had accumulated sufficient information to generate processes independent of environmental conditions and consumed food. Life can resist undesirable changes in the environment, returning the latter to the state optimal for itself. Clearly, these impacts based on genetic information accumulated in the memory of organisms cannot be described with the same equations as the processes of physical self-organization and self-organized critical states – the entire information of which is contained in the environment and external energy fluxes. Specifically these equations cannot be applied to the processes of life evolution connected with further change and accumulation of genetic information.*

The information flux coming to biota with nutrients does not affect in any way the accumulated information and, therefore, the latter remains unchanged for a time period less than the period of evolution. In this period the information only gets erased (decay). In this sense, life behaves like a physically closed system.

Although life is an open system, the energy (nutrient) fluxes consumed by it, as well as any other impacts of the environment, are of such low order that they cannot raise the level of life ordering (i.e., information supply). *Therefore, in living systems an analogue to the Second Law of Thermodynamics is functioning – during a time period much less than the period of evolution, living systems can lose information, that is, entropy can increase despite the consumption of the external energy flux.*

So, biota is the basic factor forming and organizing the geographical environment, and this environment (see Section 2.1), together with the solar energy flux, determines and governs the processes of physical self-organization and organized critical states.

2.3 REGULATION OF THE ENVIRONMENT OR ADAPTATION TO IT?

As was mentioned above, biota regulates, forms, and organizes the environment. Examples of life functioning on the Earth were given in Section 1.4. The following material follows the publication by Gorshkov et al. (1999).

The climatic postulate forming the basis of neo-Darwinism and of favourable climate for life on the Earth due to its fortunate position orbiting the Sun, should be revised, as analysis of planetary climate stability has shown (Gorshkov et al., 2000; Gorshkov et al., 2001). In fact, the climate of our planet is unstable and can only be

Table 2.1. The present and physically stable state of the Earth.

Gorshkov et al. (2000).

State of the Earth	Solar constant (W m^{-2})	Albedo (%)	Average surface temperature (°C)	Possibility of life existence
Present Earth	1,367	30	+15	Possible
"White" Earth	1,367	80	−85	Impossible
"Hot" Earth	1,367	75	+400	Impossible

in one of two stable conditions – "white" Earth completely iced, and "hot" Earth with the oceans evaporated (Table 2.1).

From the data of the paleotemperature, in the process of evolution the Earth had approached the edge of transition to physically stable conditions: in the Archaean era its average temperature had risen to 50°, from some data, to 90°C (Sorokhtin and Ushakov, 1989), and in the period of a first large-scale glaciation of the planet it dropped to 5°C. Physically, transitions into these stable states are free to take place. *Therefore the observed environmental stability for life can only be explained by the regulating function of biota* (Figure 2.1).

According to the concept of neo-Darwinism, the environment of our planet was originally favourable for life because of its fortunate position in orbit around the Sun. In these conditions, biota has adapted to practically all changes of the environment. Respectively, during 4 Gyr there have been no catastrophic events that biota have not adapted to. Changes in the environment can also take place due to the

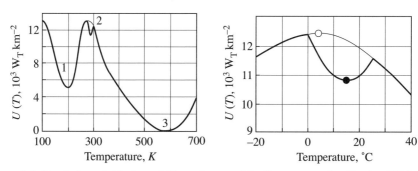

Figure 2.1 Physical and biotic stability of potential climates on the Earth: $U(T)$ is the potential function of Lyapunov (left diagram and the upper part of the right diagram). The thick line is the biotic feature of the function behaviour determining the stability of the present climate. The thin line is the thermodynamic behaviour without any special features and corresponds to unstable equilibrium. The filled circle is the mean global air surface temperature for the observed climate. The open circle is the temperature of unstable equilibrium on the lifeless Earth or after the violated biotic regulation of the environment. 1 – the state of complete glaciation ("white" Earth), 2 – present state, 3 – the state of complete evaporation of water ("hot" Earth).

Gorshkov et al. (1999).

functioning of biota itself. Biota also adapt's to these changes. The basic property of life within this concept is the evolution of life through continuous adaptation to external conditions. There are no chosen optimal conditions for life, and the genetic adaptation is provided by intraspecies variability of the genome. Any genotypes, the owners of which produce most of the surviving progeny, take root in the population. Any types of organisms capable of profusely producing can form part of the Earth's biota.

Within this concept, an important observed phenomenon is ignored – ecological limitations of the population density for most natural species. It does not explain motives for the formation of strictly correlated communities of species, why these communities and their environment are stable, and why the average lifetime of species in different biological kingdoms is approximately equal. Also, there is no explanation for the following two empirical facts: (1) why, despite the environmental changes, especially due to biota itself, the environmental conditions do not exceed the limits of life existence; and (2) why, despite a continuous adaptation, all species remain strictly discrete, and there are no transition forms either between present species simultaneously existing in the biosphere or between species in the geological past (by evidence of palaeontologic data).

The concept of biotic regulation of the environment removes these contradictions. *The main property of life is the ability of species to perform certain functions on maintaining the environmental conditions chosen for life.* Species maintaining these environmental conditions should preserve their genetic programme unchanged and cannot continually adapt to any fluctuations in external conditions. They should have a mechanism, which would stabilize the genetic programme and prevent an accumulation of mutational substitutes on geological timescales. The discrete character of a species proves the absence of a mechanism for continuous adaptation. The lifetime of a species, from palaeontological data, averages millions of years.

The observed genetic variability of natural species corresponds to accidental deviations from the normal genetic programme, and not to a recorded stabilizing selection. The stabilizing selection is a mechanism to select species with considerable accumulated genetic deviations and to eliminate them from their population. The band of accidental genetic deviations from the normal genome not recorded by the stabilizing selection is a genetic polymorphism within which all species of a population have the normal genome and a similar level of competitive ability (see for details Gorshkov, 1995; Gorshkov et al., 2001). The evolutionary transitions to new species are limited and take place only if their ability to stabilize the environment in ecological communities is preserved.

During long geological time periods of hundreds of millions or billions of years, the environment can change drastically due to geophysical or cosmic factors. In such cases, biota completely transforms itself and its structure, forcing out previous species and communities of organisms. Such a transformation can be exemplified by the transition from the reduced to oxidized environment and by an appearance of photosynthesizing organisms with a stronger competitive ability than anaerobic organisms. But in the period of transition the environment had been controlled by biota with subsequent forcing out of old (anaerobic) organisms to marginal ecological niches. With not so great changes, the dominant organisms can remain in communities, and others can be replaced with new ones.

The complicated interaction of biota with the environment leads necessarily to the formation of communities of species that interact with each other and with the environment (i.e., to a structuring of life and the environment). This process is similar to the correlation of the cells and organs within one multicellular organism. However, an important specific feature is that within communities, species of one kind interact with species of another kind. Only those species exist in the biosphere and form communities, which regulate the environment, maintaining an optimal population density. They regulate the environment and prevent an explosive growth in population. If such a growth does take place, it leads to a decrease of the normal work on environmental stabilization, and it is similar to the cancerous growth in a multicellular organism. The community organisms and their genome constitute a single system, whose function is aimed at maintaining the stability of this system and providing an optimal participation of each species in regulation of the environment. This is similar to functioning of the cells in a multicellular organism where they work not for themselves but for the whole strictly organized system of the organism. Thus, the work of the community is based on interaction of the species forming this community, but specimens consisting of these species work on the basis of competitive interaction to maintain the stability of the normal genetic programme (Gorshkov et al., 2001). Stabilizing the environment, the correlated communities of organisms compete with similar quasihomogeneous communities. The competitive interaction is aimed at forcing out of the population of the communities, those communities which cannot regulate the environment. Communities are exclusively complicated systems with a common genome, which is also characterized by genetic polymorphism. Therefore, communities in a population are not identical, but have some differences.

The concept of adaptation does not require the structuring of biota, on the contrary, it is a chaotic set of species in which the organisms with maximum productivity force out other species, gradually transforming into new species. In contrast to this point of view, the concept of biotic regulation of the environment presumes the structuring of biota into correlated communities, which regulate the environment in a competitive interaction with similar communities, and on this basis ensure the stability of these communities.

Biota cannot be a globally correlated system (Lovelock, 1982). In the absence of competitive interaction, it is impossible in such a system to distinguish between progressive changes and regressive ones, which are equiprobable within it. Therefore changes in such a system will weaken its regulated nature (i.e., lead to degradation). This process is equivalent to an accumulation of irregularity (entropy) in closed systems (see Section 2.2). With the assumed existence of such a system of life, it would not be able to evolve, because all changes in it should be prohibited. *Only one mechanism is capable of preserving the regularity and at the same time ensuring evolution (i.e., a stabilizing selection that acts by means of the competitive interaction of the totality of homogeneous communities)*.

Thus biota is the main actor in the geographical environment and the basic mechanism ensuring the regulation and dynamic stability of the environment, including the climatic stability. Preservation of natural biota in sufficient volume is the key problem of preservation of life and civilization stability (i.e., the key problem of ecology).

3

Structural units of the geographical environment

3.1 LANDSCAPES AND ECOSYSTEMS

As far back as the last century the notion of "landscape" originated in geography, leading to the scientific discipline of "landscape science". An appearance of landscape as a structural unit of the geographical environment was quite natural. As A. Isachenko wrote (quoted in Dronin, 1999): "New geography has resulted not from academic discussions, but it appeared in the process of practical studies of the nature of Russia, and it was introduced by naturalists who did not even consider themselves geographers." They have also brought forth the concept of landscape. The appearance of the notion of landscape and then of landscape science is connected with the fact that in a real geographical environment landscape structural units are clearly seen in principally vegetation structural units. Apparently, for this reason the soil scientists V. V. Dokuchaev and his pupils – geobotanists T. G. Tanfilyev and G. N. Vysotsky and the then geographer and biologist L. S. Berg – actively worked on the problems of landscape science.

A similar process also took place in the field of biology, where the term "ecosystem" appeared. Biologists have gradually begun to select ecosystems as territorial units, differing, first of all, by vegetation, rejecting the idea of a "dimensionless" ecosystem. Thus, the geographical environment (the same as biosphere and environment) is really structured, with two important sciences studying these structures.

The most widely used terms to denote the structural spatial units are *landscape* in geography (at present the term "geosystem" is used), and *ecosystem* in ecology. These notions have much in common: first, they denote a site as the Earth's surface; second, they include both abiotic and biotic constituents; third, all components within them are interconnected and, being mutually dependent, form an integrated system; and fourth, they are the result of evolution.

Landscape and ecosystem also have a feature in common: their boundaries and scales are uncertain. So, of eight definitions of landscape, Reimers (1990) mentioned only one, where its boundaries are defined as "natural borders". In the explanatory dictionary *Protection of Landscapes* (1982) nothing is said about the boundaries in the definition of landscape, and the definition of the term itself, "landscape boundary", does not help their determination either. An explanation of landscape is too broad: from the whole Earth to very small sites (a territorial system that consists of interacting natural or natural–anthropogenic complexes and components of a lower rank). A similar interpretation of landscape is given in the following definition: "a totality of interconnected and mutually dependent natural objects and phenomena, historically forming a physico-geographical complex or a number of complexes developing in time" (Reimers). The term "elementary landscape" introduced in *Protection of Landscapes* (1982) adds an uncertainty in determination of landscape boundaries and, hence, a possibility of their subjective selection.

Solntsev gave a most precise definition of landscape. Later on it formed the basis of the definition in the *Encyclopaedia Dictionary of Geographical Terms* (1968) of: "a concrete territory homogeneous in its origin and indivisible by zonal and non-zonal indications, with a single geographic foundation, relief of the same type, common climate, uniform combination of the hydrothermal conditions of soil, biocenoses, and, hence, a set of simple geosystems (facies and tracts) of similar character."

The same polysemy also refers to the term ecosystem, which can also be applied to "a drop of water with micro-organisms contained in it and in the biosphere, on the whole" (Dediu, 1989; Reimers, 1990). In the definition suggested by the international standard (ISO G17-3, 1993), the ecosystem is also understood rather in a broad sense: "a correlated system of living organisms and their environment in which a cyclic exchange of substance and energy takes place." Uncertainties in explanation of the term ecosystem are, of course, connected with the uncertainty of understanding the term ecology.

Rozenberg (1999) performed an analysis of definitions of the term ecology. More than 60 definitions were analysed, which revealed diverse definities ranging from biology to social sciences: "its (ecology) boundaries go beyond the limits of even the synthetic biological discipline (they say about 'social ecology', 'engineering ecology', 'political ecology', 'cultural ecology', etc.)". From this follows the same diversity of ideas about the term ecosystem.

In both these objects (landscape and ecosystem) a desire is observed to emphasize their elementary indivisible constituents in a space called landscape or ecosystem. The least unit of landscape is termed "facies" – a site over which the homogeneous lithologic composition remains the same as well as the character of its relief or microrelief, microclimate, and one biocenosis (*Protection of Landscapes*, 1982; Reimers, 1990). In ecology, the notion of facies is also used as the least phytocenologically distinguishable unit of a vegetation community, in which some species prevails; a combination of similar biocenoses. It is seen from this explanation that a biological facies is not a geographical facies, since it contains several biocenoses, whereas the latter contains one biocenosis.

Several definitions for every term testify to the absence of any generally accepted theory, within the framework of which a single definition of the term is only possible.

These similarities motivated many authors to identify landscape with ecosystem and vice versa (Polynov, 1982; Tiuriukanov and Fedorov, 1996). Landscape scientists, nevertheless, believe that there are differences between them. "In the ecosystem an analysis of the bonds has a biocentric aspect: an effect of all factors on living components and on the host of the ecosystem is being studied ...; in the consideration of natural landscapes and natural geosystems this is not stressed; elements and their bonds are considered equivalent ..." (Polynov, 1982). This suggests that landscape and ecosystem hold similar territorial structures with geographers and ecologists (biologists) having slightly different views over the similarities. The statement is incorrect to suggest that ecologists only consider the effect of the environmental factors on biota, whereas, in fact, they also study the effect of biota on the environment. A problem arises whether the postulate is correct that the elements and the bonds in a given space are equivalent. It is also of interest that in the definition of "anthropogenic landscape" biotic "host" appears – a human being – as the main and leading element of a given space (Polynov, 1982). Why then cannot biota be such a leading element in the natural landscape? Biota is such an element. To substantiate this conclusion, it is enough to compare the power of biota, the power of transpiration, and the power of a civilization economy (Table 1.2).

Attempts to separate the notions of landscape and ecosystem seem unconvincing: these are simply views of the same structural unit of the biosphere or geographical environment (envelope) taken by specialists in different spheres, in which the basic function of these units is not considered. This is explained by the fact that geographers and biologists underestimate the role of biota in the formation, organization, and regulation of the environment (see Section 1.1) (Kondratyev et al., 2001b; Losev et al., 2001).

Section 2.2 contains examples of the formation by biota of its own surroundings – the geographic environment and its components. It is seen from Table 1.2 that the power of biota is an order of magnitude greater than the power of civilization, which has changed the planetary land surface by 63% during 10,000 years. Why then cannot biota change the environment during 4 Gyr, having such huge power?

To answer this and other questions, consider the elementary structures of landscape and ecosystem, which do form units of space – the geographical environment or the biosphere.

3.2 FACIES AND BIOGEOCENOSIS

The definition of facies in landscape science was given above. It is interpreted as an elementary, further inseparable part of landscape. The notion of biogeocenosis used in ecology and suggested by Sukachev (1964) as "a totality of homogeneous phenomena (atmosphere, rocks, vegetation, animals, micro-organisms, soil, and hydrological conditions) over the whole Earth's surface, with special interactions between these constituents and a certain type of exchange of matter and energy between them and with other natural phenomena, and representing an internal contradictory dialectic unity, constantly developing" erases differences between a landscape facies and biogeocenosis (Polynov, 1982).

Biogeocenosis is often considered a synonym of ecosystem, but the definition by Sukachev (1964) does not suggest such a conclusion should be reached. Biogeocenosis is often considered as an elementary component of ecosystems. In 1968, Timofeev-Resovsky (quoted here from his inaccessible writings mentioned in Tiuriukanov and Fedorov, 1996) wrote that "V. N. Sukachev was correct to detect in the biosphere of the Earth more or less discrete sub-units separated from each other by various boundaries – biogeocenoses. *Biogeocenoses are elementary structural units of the biosphere and, at the same time, an elementary unit of the biological cycle, that is, biogeochemical work taking place in the biosphere.*" Reimers (1990) similarly notes, biogeocenosis is an "elementary biochorological unit of the biosphere." Comparing the definitions of the terms facies and biogeocenosis we see that they are identical. Timofeev-Resovsky (1969) noted this, pointing out that landscapes are greater complexes of biogeocenoses (Tiuriukanov and Fedorov, 1996).

The Russian landscape scientist Isachenko (2002) also considers the identity of biogeocenosis and facies indisputable. Sochava (1978) considered it worthwhile to preserve both terms in landscape science, believing biogeocenosis to be an elementary geosystem, and facies – a combination of homogeneous biogeocenoses.

Both facies and biogeocenoses have the same problem as do landscapes and ecosystems – the identification of their boundaries. Proceeding from definitions of these elementary structures, it is impossible to determine precisely the boundaries of facies and biogeocenosis. Therefore, biogeocenosis is often combined with an ecosystem, and facies with a tract, locality, or even landscape. It is often observed in the field of landscape science. Apparently, the use of the terms depends on the scale of studies.

Nevertheless, as in the case of definitions of landscape and ecosystem, the landscape scientists do try to emphasize the difference between the terms facies and biogeocenosis, believing that the latter term points to biocentricity, and in facies all the components are equal. As a result, the bonds between these components are equally considered, whereas in biogeocenosis their impact on the life system is studied. In fact, in biogeocenosis the impacts of biota on all the environmental components are studied. It is life (biota) that forms facies and landscapes and biogeocenoses and ecosystems that form the structural units of the biosphere or the geographical environment.

In reality, if we compare, for instance, the volume of the run-off into the ocean of suspended and dissolved substances ($21\,\mathrm{Gt\,yr^{-1}}$) with the mass developed on land in the process of photosynthesis of organic matter (more than $1,000\,\mathrm{Gt\,yr^{-1}}$), we see that the latter flux is several orders of magnitude greater than the former one. But it should be borne in mind that 80–90% of the weathered material consisting of alluviums and dissolved substances are also the result of the work of plants, micro-organisms and fungi, proved experimentally by Lapo (1987).

The mass of volcanic substance annually coming to the planetary surface is about 1,000 times less than the global mass of organic production (Losev, 1989). A comparison of the volume of the annual production in the form of organic carbon with a pure flux of inorganic carbon from the Earth due to degassing shows the inorganic carbon flux is 100,000 times less intensive. Finally, the land biota provides 70% of the continental moisture cycle due to transpiration. These facts testify to a much greater intensity of the biological cycle of matter (the difference is several orders of magnitude) than the geological and geophysical cycles taking place on various scales, as well as in the facies of landscapes and biogeocenoses. Therefore, various interconnections between the components of these structures cannot be considered equal. First of all, one should bear in mind the existence of life within them and biotic activity (Kondratyev et al., 2001a).

3.3 ELEMENTARY UNITS OF THE BIOLOGICAL CYCLE

It was mentioned above that Timofeev-Resovsky defined biogeocenosis as an elementary unit of the biochemical cycle. He wrote: "the normally functioning biosphere of the Earth not only provides humankind with food and most precious organic material, but also *maintains an equilibrium state of the gas composition of the atmosphere and the solutions of natural waters. Hence, a violation (quantitative and qualitative) of the biospheric function by a human being will not only reduce the produce of organic matter on the Earth, but also violate the chemical equilibrium in the atmosphere and natural waters*" (cited in Tiuriukanov and Fedorov, 1996). In 1968 he noted that "... the biosphere of the Earth – a gigantic living factory, which transforms the energy and substance on the surface of our planet – forms both the equilibrium composition of the atmosphere and the composition of solutions in natural waters, *and through the atmosphere it forms the energy budget of our planet*. It also affects the climate. One should remember an important role in the global water cycle of evaporation of water over vegetation cover. *Hence, the biosphere of the Earth forms the human environment. And the careless attitude to it, violation of its correct functioning will mean not only a violation of food resources for humans and a number of raw materials necessary for people but also a violation of the gas and water environment of humans. Finally, people without the biosphere or with a poorly functioning biosphere cannot exist on the Earth at all*" (Tiuriukanov and Fedorov, 1996). In such a way the outstanding Russian scientists briefly formulated the idea of biotic regulation of the environment.

Biogeocenoses are elementary cells of this regulation. Timofeev-Resovsky emphasized the necessity of their studies: "most of biogeocenoses in a state of *prolonged dynamic equilibrium are very complicated self-regulating systems*. Therefore it is most important to study causes, mechanisms and conditions to maintain this dynamic equilibrium in biogeocenoses" (cited in Tiuriukanov and Fedorov, 1996).

Vernadsky had expressed similar ideas: "an organism is an integral part of the Earth's crust, its produce, part of its chemical mechanism.... Life changes conditions of chemical equilibrium for a substance, and for all chemicals constituting this substance" (cited in Tiuriukanov and Fedorov, 1996).

Gorshkov further developed the conceptual ideas of Vernadsky and Timofeev-Resovsky concerning the biotic regulation of the environment. He generalized a huge volume of material accumulated by geographic–geological and biological sciences in the 20th century (Gorshkov, 1995; Gorshkov et al., 2000). As a result, a theory of biotic regulation of the environment has been developed. But it does not follow the "Gaia hypothesis" by Lovelock (1982), who could not explain a real mechanism of regulation and stabilization of the environment by biota. The concept of biotic regulation was briefly discussed above (Chapter 2). For more details see the monographs of Gorshkov (1995), Gorshkov et al. (2000), and also Gorshkov et al. (1999).

The theory of biotic regulation explains the causes of the formation of biogeocenoses (in the terms suggested by Sukachev (1964) and Timofeev-Resovsky) or natural communities of organisms – elementary ecosystems (in the terms used by Gorshkov (1995)), their stability, limiting the population size of species, and the stability of the habitat (i.e., the aspects that should be studied), as called for by Timofeev-Resovsky.

Biota cannot be a globally correlated system, Lovelock (1982) (see Section 2.3), since in a regulated and internally correlated system any processes should lead to a violation of this order. In fact, biota is organized into small regulated and correlated communities of organisms – biogeocenoses – which are elementary units of the biochemical cycle and environmental regulation. In each elementary unit the biochemical cycle is rigidly closed. The degree to which the cycle is closed is confirmed by the statistical law of great numbers, according to which a relative fluctuation in the system of independent elements is determined by the value $1/N^{1/2}$, where N is the number of non-correlated parts of the system. In biogeocenosis, both synthesis and decomposition of organic material are accomplished by a great number (hundreds of thousands) of independent elements. For instance, synthesis in a tree is realized through needles or leaves (which average 200,000 per tree), which act independently of each other and even compete for the solar radiation flux. Decomposition is realized by micro-organisms (billions of species) and by fungi hyphae in soils.

Because of huge amounts of micro-organisms and fungi hyphae in soils, the organic decomposition is characterized by low natural fluctuations and, as a rule, does not introduce any disturbance into biogeocenosis (local ecosystem). But usually natural systems include such large moving animals like elk or deer consuming vegetative organic material and seemingly breaking the closed cycle of nutrients. One way to prevent an irreversible increase of openness is to reduce the share of consumption by these animals. Experimental data on numerous natural ecosystems (landscapes) show that the share of consumption by large animals does not exceed 1% of pure primary production (Figure 3.1). The feeding territory of large animals includes a multitude of biogeocenoses, and the density of their mass per unit area of the ecosystem (landscape) is constant through a time period of several years. In the mid-latitude forests it constitutes 2–3 kg ha^{-1}.

Correlation of organisms (producers and consumers) in the community is provided by the flux of substances and energy. It is at a maximum when the size

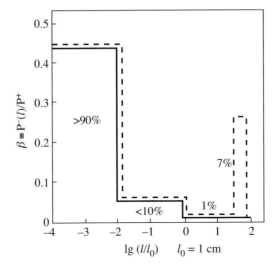

Figure 3.1 Distribution of the rate of decomposition (destruction) of organic substances on land by the size of organisms (bacteria, fungi, insect, and animals). $P^-(l)$ is the spectral density of relative destruction produced by the organisms of size L; P^+ is the land vegetation produce (net primary production). The solid line is the universal distribution observed for every non-perturbed ecosystem. The area under this line is equal to unity. The dashed line describes the present distribution over land, corresponding to the antropogenic disturbance of land biota. The area under the antropogenic peak (7%) corresponds to human food, stockbreeding, and wood consumption. The difference between the areas under solid and dashed lines characterizes the openness of the biochemical cycle.

Compiled from data on the global carbon cycle (Gorshkov, 1995).

of biogeocenoses (communities) is minimum. The size of biogeocenoses ranges between 1 cm and dozens of metres (i.e., not more than the canopy projection of higher plants in the forests). An indirect characteristic of the size (boundaries) of biogeocenosis is the "length of saturation" within which the species consuming most of the energy flux of the community are located. Usually the length of saturation of the number of species is supposed to be independent of the share of energy consumed by them, which leads to strongly overestimated boundaries. The latter can be determined experimentally by measuring the space of circulation of radioactive marks of non-volatile elements introduced into biogeocenosis without destruction of its natural structure (Gorshkov, 1995). Biogeocenosis is indivisible since, when destroyed, it loses its basic function.

Quasihomogeneous, correlated communities – biogeocenoses – form populations (ecosystems). The competitive interaction of biogeocenoses in them provides the elimination of those biogeocenoses, which can no longer regulate and stabilize the environment, since biogeocenoses, like any correlated system, can disintegrate.

Thus within the framework of the concept of biotic regulation of the environment, the notions biogeocenosis and ecosystem are clearly defined and have physically determinable boundaries and functions. *An ecosystem, as mentioned*

above, is a population of quasihomogeneous, further indivisible elementary cells –
biogeocenoses. Apparently, if facies and biogeocenoses are identical, then a set of
quasihomogeneous facies forms a landscape, since the same primary structural
elements of the environment should form similar (identical) structures of a higher
rank. A set of heterogeneous ecosystems can be called a territorial or, if it is more
spacious, a regional ecosystem, and when united, will form a global ecosystem – the
biosphere. *Thus the biosphere should be defined as biota and the environment interact-*
ing with it. Biota over a given territory is the sum of all natural communities of
organisms (i.e., biocenoses, which regulate and stabilize their environment). It does
not contain cultivated plants and animals.

Biogeocenoses and ecosystems form, regulate, and stabilize the environment,
therefore, it is important to mainly study them from this point of view. Timofeev-
Resovsky (1968) noted that these structures form the gas composition of the atmosphere and chemical composition of waters, affecting the climate and forming soils
(i.e., they represent a biological reactor in which the biological and geological cycles
of substances interact: Dobrovolsky, 1999). In other words, they form all the abiotic
components. Therefore, the latter cannot be studied without relevant connection
with the biota forming them, like the biota of biogeocenosis without its interaction
with the inert components.

3.4   LANDSCAPES AND ECOSYSTEMS, FACIES, AND
BIOGEOCENOSES: CONSEQUENCES AND CONCLUSIONS

From the theory of the biotic regulation of the environment, let us follow a number
of important consequences for landscape science.

Biogeocenoses, elementary, further indivisible cells of the cycle of nutrients,
existing objectively, are similar to facies – lowest in hierarchy, further indivisible
units of landscape. Their division, elimination, or severe destruction lead to the loss
of their main function – regulation and stabilization of the environment. It is for this
reason that they are indivisible. If this were not so, then it would be necessary to
show what determines the indivisibility of facies.

Between ecosystems (landscapes) and biogeocenoses (facies) there are no intermediate, objectively existing, spatial systems, whereas now in landscape science there
are tracts and localities between facies and landscapes. For theoretical and practical
purposes, this is possible with precise, generally accepted grounds for their selection,
with an account of the main function of facies (biogeocenosis). In other words, their
boundaries should not split biogeocenoses (facies) and ecosystems (populations of
quasihomogeneous biogeocenoses).

The territories of landscapes meeting the requirements of integrity and homo-
geneity (as mentioned in *Protection of Landscape*, 1982) should coincide with the
ecosystem. As a matter of fact, they are synonyms. It turned out that the two most
important sciences have a language of their own to denote the same structural
natural units.

All this refers to natural ecosystems and natural landscapes. As for the notions "agroecosystems", "urban", "technogenic", and "anthropogenic" ecosystems and similar landscapes, the theory of the biotic regulation of the environment also suggests relevant important conclusions.

Between natural landscapes and natural ecosystems, on the one hand, and their anthropogenic versions, on the other hand, they have nothing in common, except that for both of them life is the dominant factor. In natural ecosystems this is natural biota organized into communities, creating optimal conditions for present and future life. In anthropogenic systems it is a human being and human communities breaking the natural mechanisms of regulation in order to build an artificial, technogenic environment of their own, while not taking care of future life, but pursuing their own, short-lived egotistic aims.

The report of the Brundtland Commission (*Our Common Future*, 1989) formulates the goal of sustainable development – provision of safe life conditions for present and future human generations. To reach it, first of all, a stable geographical environment is needed. In anthropogenic systems the basic process – formation of a stable environment based on natural cycles of nutrients – ceases upon reaching a certain limit of disturbance (destruction) of the elementary cells of this mechanism – biogeocenoses (facies) or, in other words, as a result of human activity going beyond the limits of the carrying (economic) capacity of ecosystems or the global ecosystem – the biosphere.

In an anthropogenic system, the environment-forming factor is human activity aimed at providing favourable conditions for the existence of only one species (*Homo sapiens*), at a given moment and in a short-range perspective. In anthropogenic systems all natural balances, processes, and regularities are actively violated. Therefore, such systems are *not* the subject of "joint creation by nature and man", as Sochava believed (*Protection of Landscape*, 1982) or coevolution of nature and society, according to Moiseyev (1995). In fact, man creates his systems on the ruins of biogeocenoses and ecosystems (facies and landscapes) when he oversteps the limits of their carrying capacity. Therefore, in the anthropogenic spatial formations there immediately appear various problems connected with environmental changes. In solving them, humans do not eliminate the basic cause of the problem, but reduce the damage caused by the developing changes, to an acceptable level established subjectively, proceeding from short-term interests and economic and financial possibilities. It means that there is nothing in common between natural and anthropogenic structures in the anthropogenic and natural landscapes (ecosystems), except the word "landscape" (ecosystem). The term landscape is a word "on loan" from the German language, where it is used only in one sense – to denote the anthropogenically changed systems, since natural landscapes in this country vanished a long time ago (see Chapters 4–6).

Similarly different (almost antagonistic) are the notions of ecosystem, agroecosystem, urban ecosystem, etc. An ecologically sustainable development, in principle, is impossible in such anthropogenic systems after exceeding the limit of disturbance (destruction) of the elementary regulators of the environment. One can only, and indeed must, reduce the level of ecological disturbances using all available technological means.

The carrying economic capacity, or the limit to disturbance of natural ecosystems (landscapes) is demonstrated in Figure 3.1 (Gorshkov, 1995; Gorshkov et al., 2000). This figure is a graphic expression of the law of distribution of energy fluxes among the consumers in biota, obtained from an empirical generalization of all available data of observations carried out in different natural ecosystems. It shows an acceptable level of consumption of net primary biological production by humans for conditions of preserved environmental stability. Humans may use in their interests not more than 1% of net primary production of global or local ecosystems. It can be recalculated into an admissible economic capacity, which will constitute 1 TW, or into a value of the mastered area, which ranges within 20–30% of land surface. The theory of biotic regulation has solved the problem which Holdgate (1994) wrote about – that on the problem of carrying capacity of ecosystems "many scientists have broken their teeth".

The above does not imply an appeal to refuse to create anthropogenic systems, which is unavoidable in conditions of our present-day civilization. But, in creating them, it is necessary to have an idea of what man is really doing, of the limits to admissible destruction of the mechanism for the biotic regulation and stabilization of the environment in each ecosystem (landscape) and in the biosphere (geographical environment) on the whole. With these limits taken into account one can speak about the possibility of ecological stability of life via a certain means of development.

In light of the above, the notion of ecology can be defined as follows: *it is the science that studies the laws and mechanisms, which provide the stability of life and environment on the Earth.* It is based on the theory of biotic regulation where the key problem is the carrying capacity of ecosystems. Humans have left behind a long traversed path, repeatedly violating local ecosystems resulting in local ecological and then resource, social, and political crises (see Chapters 4–6), before in the 20th century they overstepped the limits of the carrying capacity of the global ecosystem (geographical environment). But still not all historians consider humankinds' development from this point of view.

4

Humans are changing the geographic environment

4.1 HUNTER-GATHERER CONQUERS THE PLANET

In the process of the evolution of biota, cardinal changes (discoveries) have taken place on the Earth, leading to the appearance of new organisms, principally differing from previous species. For example, an initial appearance of multicellular organisms, leading to vertebrates and animals with a constant body temperature (birds and mammals). Such changes are very rare phenomena. There have been several changes over the last 4 Gyr. An appearance of human beings is also connected with a fundamental, genetically fixed change, which has put human beings in an exclusive position within the animal kingdom.

All known organisms, except human beings, contain in their genome, information needed for their existence, which is inherited (see also Section 2.3). In the animal kingdom, when parents care for prodigy, a hereditary programme of teaching is possible. Moving animals can be taught and trained, but this information cannot be inherited. All extragenetic memorized information lasts only for the life of one specimen. This prevents the transfer of incorrect or unnecessary information to following generations (Gorshkov, 1995).

Some animals use tools: vultures break eggs with stones; the Darwinian bird on the Galapagos Islands uses thorns to extract insects from cracks in trees; chimpanzees use sticks to feed on termites. Growth of young specimens of these animals away from their parents has shown that these skills are hereditary (i.e., they are contained in the genome of a species). In other cases, visual teaching methods are possible. For instance, in England, birds can open milk bottles by pecking through the foil top and demonstrate this method to other birds within their flock. Monkeys sometimes wash sand off fruit. This is not an element of species culture, but is a genetically programmed limit of the ability to acquire and process information in the extragenetic memory of these species. The loss of this information does not threaten the survival of the species. In changed conditions they lose this information but, in returning to normal conditions, they again acquire the respective skills (Gorshkov, 1995).

An appearance of culture accumulated and passed over from generation to genera-tion in the process of teaching and not contained in the genome of a human being, is a unique human quality. The cultural heritage is based on the genetically fixed behaviour. It is only human beings who have learned to use scientific, technical, and other extragenetic information supplies – cultural heritage – to raise the species competitive ability and to change the environment. This transition resulted from a divergence of a single species-precursor in whose genome a cardinal progres-sive reorganization had occurred. It had led to an evolution of the genome and morphological structure of a human being towards a better use of his abilities. As a result, the human brain had grown in size in tandem with his capacity for the generation and accumulation of cultural information and the appearance of language as a means to to communicate this information.

The process of formation of the human genome had taken place until the rates of accumulation of genetic and cultural information had become comparable (i.e., until morphological changes and cultural skills had equally increased the competitive ability of the species). But as soon as the rate of accumulation of cultural informa-tion exceeded the rate of evolution (accumulation of genetic information), the species stopped changing genetically. Probably, this occurred at the stage of the appearance of Cro-Magnon man not later than 100,000 years ago (Gorshkov, 1995).

Ancient Hominidae, precursors of the present *Homo sapiens*, had been gatherers. Their remains found in Kenya and South Africa testify to their ability to climb trees. But they also could have walked on two legs and probably had extracted plants, roots, and grains. At the same time, they may have fed on insects, their larvae, and small animals, varying their vegetarian diet. As other large herbivorous, Hominidae had been minor disturbers of the environment, performing an important function of maintenance in a "working state" of the genetic programme of regulation and stabilization of the environment by communities of organisms – biogeocenoses.

First tools of ancient pre-*sapiens* had appeared to be used to gather vegetable foods. The next developmental step was to feed on the corpses of animals killed by predators. Males of the first *Hominidae* may well have sporadically hunted for small animals. Probably, the first division of labour had already taken place: females remained gatherers and males gradually transformed into hunters and fishermen.

An increase of diversity in food sources, mastery of fire, improving and making new tools, and provision of a means of protection from bad weather had created conditions for the human pioneering of new landscapes and climatic zones with seasonal changes of vegetation. These had been prerequisites for populating and developing the planet. Among the reasons for humans' propagation could have been changes of their initial habitat, growth of the size of population, division of groups, and, of course, deliberate actions connected with human creative abilities.

Africa is considered the cradle of humankind, but there is evidence, though still not very convincing, of a possibility of the existence of an East Asiatic centre for the origin of humankind. Numerous archaeological data testify to the fact that pre-*sapiens* started settling approximately 1.8 Myr ago. First settlers had been

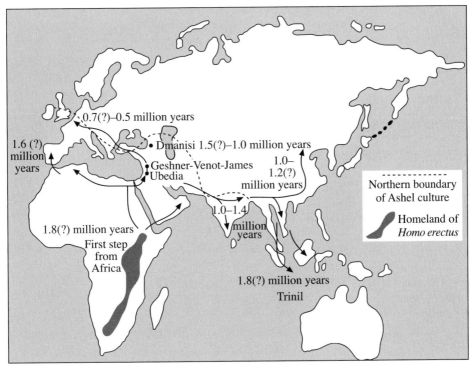

Figure 4.1 Propagation of *Homo erectus* to North Africa and Eurasia.
Velichko and Soffer (1997).

Hominidae and then the modern *Homo sapiens*. Population of humans over all the continents had trigged the process of globalization of this species. Human beings had become a planetary phenomenon.

Colonization of the Earth by *Hominidae* continued during almost the whole Paleolithic period and took 99% of the time taken from appearance of pre-*sapiens Hominidae* to modern *Homo sapiens*. This slow process had been started by the precursor of *Homo sapiens* – *Homo erectus*. From East Africa, he first moved to North Africa, then through the Lybian corridor – to Eurasia – and farther to the east (Figure 4.1). The average rate of movement when populating new areas did not exceed 1 km yr^{-1} (Velichko and Soffer, 1997). The mechanism of populating remains poorly studied, but two ways were most probable: linear and diffuse frontal as a combination of both. Propagation from Africa to Eurasia of primitive humans could most likely have been linear, since propagation had taken place via land connections of the Arabian peninsular with Africa (Lybian corridor). Also, it could have been realized through the regions of the Strait of Gibraltar and Strait of Sicily with the growing glaciation of the Earth and the accompanying lowering of sea level. On the whole, the process of population is characterized by strong irregularity in time and space. Settling in new geographical zones was followed by death for large groups of humans, which stopped the process of populating for some time, with a repeated settling taking place after. These deaths may have been connected with both natural disasters and outbreaks of hunger and diseases.

This slow and uninterrupted process had exhibited some regularities. From Africa, the settling had been realized through narrow isthmuses northward, and then a gradual fan-like propagation had begun ending with populating both high-latitude Asia and then through the tropical zone of Indonesia and New Guinea to the southern hemisphere – to Australia. Australia was populated, from different estimates, during the last 40,000 to 60,000 years. The movement was from the north to the south. Humans moved very rapidly across the whole continent and reached Tasmania island 35,000 years ago (Velichko and Soffer, 1997).

The pioneering of North and South America was similar to that of Australia. At present there is a dominant hypothesis of penetration of first settlers to this continent through the Bering isthmus, as a result of a lowered sea level due to glaciation in the northern hemisphere. Based upon dated archaeological findings, this happened not more than 30,000 years ago (possibly later, but not less than 11,200 years ago). It took about 2,000 years to move from the Bering Strait to Patagonia. However, some scientists believe that South America could have been populated by sea. Development of Australia, northern Siberia and both Americas shows that the process of development of the planet by humans accelerated at the close of the Paleolithic period, and the rate of populating new territories sharply increased by the end of this period.

Of course, without mastering fire, human beings could not populate vast spaces of middle and high latitudes. Precise timing of this event has not been determined, but findings in Kenya (East Africa) and China testify it was a long time ago.

Traces of the use of fire

When studying camp Chasovandja in Kenya, a joint occurrence was found of split bones of animals, rough stone tools, and small lumps of burnt clay. Study of these lumps showed that clay had been burnt at a temperature not higher than 400°C. As a rule, forest and steppe fires give higher temperatures, therefore these lumps are considered traces of a most ancient campfire lit by man. In a carst cove at Cotzetang near Beijing in which ancient human beings had lived, a 7-m layer of compressed ashes was found. Ancient humans living in this cove had kept up a fire, since they had not known how to make fire. This campfire was lit 400,000 to 500,000 years ago (Bondarev, 1998).

During the last stages of global population the modern human had accelerated the development of the planet. Having trained his speech, he had taken priority over the Neanderthal man and forced him out. Before speech training, all the extragenetic cultural information ought to have remained in the brain of each human, which would have led to a rapid growth of brain volume. This was at a maximum with

Neanderthal man who lived 20,000 to 40,000 years ago. His speech had not been trained sufficiently and, therefore, an accumulation and reporting of information from generation to generation had been realized mainly by mimicry and gestures.

Speech is a cultural language of sound signals transmitted to the next generations via the process of teaching, which sharply differs from the genetically fixed signals of other animals. Teaching has become a most important element in the development and formation of new generations, which has fundamentally changed the relationship between generations (parents and children) compared with other animals. Having learned to speak, modern humans finally mastered the land much more rapidly than did *Homo erectus*.

Specialization of animals is genetically fixed, whereas with human beings it is connected with accumulation, formation, and transmission of extragenetic information. Now some individuals could use and transmit specialized information. All cultural information, continually growing as humankind developed, gradually became fixed in the brains of some members of the community (but not in the brain of each human being) (i.e., a kind of cultural specialization began to form). This led to the necessity to form a new social structure, never before seen in biota). Such a structure was based on cultural specialization with a need for interaction between members of the community, not only relatives. It was an internally correlated system (Gorshkov, 1995). Rudiments of such a community had appeared at the time of hunter-gatherers. Initially, it was a division between males as fishermen and hunters, and females as gatherers. Then emerged individuals who could not only maintain a fire but could also make fire from scratch. Then, the manufacturers of stone tools appeared, etc.

The process of conquering the globe by hunter-gatherers had required new technologies. To preserve food, to transport food and tools, relevant capacities were needed; the body ought to be protected, humans ought to be sheltered from bad weather, rains, frost, etc. Cro-Magnon man – the type of present human appearing about 100,000 years ago – was able to use these respective technologies.

During the Paleolithic and Neolithic period (between 2 Myr and 12,000 years ago) humans populated the whole planet. Material culture of the Paleolithic period had been common for all of humankind in the past. Means of obtaining food had been common, too: gathering, hunting, and fishing. All this points to the unity of the origin of humankind and the regions from where the population of the planet had started.

Conquering the planet is one of the greatest events in the Earth's history, a turning point in the evolution of life medium – the geographical environment (biosphere). It was also connected with a new process appearing first in the sphere of biota. First, the behaviour of any animal had been determined by the environment (including other organisms) and by genetic programmes within rigidly correlated communities. Now a living being appeared who, using his new ability, began to depart from this determinancy and adjust the environment to the needs of his own. Many authors call it "adaptation" or "interdeterminate adaptation". In fact, it had been an adjustment by man of the environment to himself and his needs, with the use of the resources of this environment. In this sense, human beings behaved similarly to biota, on the whole, in that they reconstructed and adjusted the geographical environment to make the best of their lives (Losev et al., 2001).

4.2   FIRST TECHNOLOGICAL REVOLUTION

Becoming a hunter involved a rapid improvement in hunting technologies, and during the mid-Paleolithic (120,000 to 140,000 years ago) a specialized means of hunting herds of large herbivorous animals was developed. Especially efficient was the hunt that involved driving a herd toward a precipice. In the well-known excavated camp of hunters of Solutre (France), bones of about 100,000 wild horses are concentrated over an area of more than 1 ha. It took 7 days for the archaeologists to excavate a 5-m layer of bones.

The mammoth was a valuable animal, which gave large quantities of meat (a male mammoth weighed 5 t). The bones and tusks could serve as a dwelling frame with a thick hide as its cover. Mammoth was a source of food for humans in North America, northern Asia, and Europe. At the close of Pleistocene and at the beginning of Holocene, many species of animals (with masses greater than 1,000 kg) had vanished. In Europe and northern Asia these included the mammoth, mastodon, woolly rhinoceros, cave bear, sabre-tooth tiger, and Irish Elk. In North America, apart from mammoth, animals included the camel, wild horse, and giant sloth.

Discussion on the causes of the "mammoth fauna" extinction

Alfred Wallace, a contemporary and fellow countryman of Charles Darwin, independently developing the theory of the origin of the species, and brought forth a hypothesis that explained the mass extinction of the "mammoth fauna" as a result of hunter activity. Discussions on what could cause such a disappearance continues to the present day. The Russian geologist V. I. Gromov (1965) connects this with climate change, while Wallace (1878) links it with the activity of primitive hunters. The Russian paleogeographer Serebrianny (1980), pointing to the fact that earlier in Earth history large animals had also died out (before humans appeared), believed that such a disappearance of the "mammoth fauna" had been a natural process which a primitive hunter may have simply accelerated.

Supporters of the hypothesis of extinction of large animals through hunting call this phenomenon the first ecological crisis on the planet or "the crisis of consumers". In our opinion, this crisis should be considered to be not of ecological origin but a resource (food) crisis, and only for those groups of hunters which had specialized in hunting large herbivorous animals. Most likely, there had been no food crisis, since ancient hunters had moved on to hunting animals of a medium size. There are also arguments favouring the natural process of extinction of large animals of the mammoth fauna. Their disappearance had taken place on all continents, with most of the species vanishing on the most thinly populated (by man) continent (America), and least of the species – on the most populated continent (Europe). Finally, study of the life of American Indians – buffalo hunters in North America – showed that while obtaining food, clothes, and building their dwellings, they had coexisted with buffalo and even further increased the herd sizes by burning the forests and enlarging the areas of prairies.

Supporters of the hypothesis of destruction of the mammoth fauna by ancient humans believe that hunger had made the hunters "discover" agriculture, which, as determined from archaeological data, originated 10,000 to 12,000 years ago. Apparently, this is not so, because the mammoth fauna had vanished mainly in middle latitudes, and agriculture appeared further to the south, in the territories populated by ancient gatherers. Gatherers had made a lot of observations and discoveries about vegetable crops: they increased the number of edible plants, discovered the medical properties of some plants, and later on they learned to obtain fibres from wild flax, kendyr, and stinging nettles, as well as learning to weave. Finally, they learned to lay food in store. It was they who would become discoverers of technologies of plant growing, not hunters and fishermen, who had lost their skill to gather. This discovery was made at the end of the Stone Age (Neolithic), and therefore, it is called the Neolithic or agricultural revolution (Arsky et al., 1997).

First centres of agriculture

The first centres of agriculture appeared in three regions of the globe – the Middle East, Egypt, and China. The heavy labour for the first crop-growers was not easy. This is testified to by excavated agricultural tools. One can imagine how difficult it was to hoe land with stone-tipped mattocks or with simple wooden sharpened sticks, to cut bundles of hard stalks of spikes with a wooden prototype of a sickle with detachable stone blades, and then to grind grain by hand on a stone slab.

The transition to brand new technologies for obtaining food was called by economists a transition from an appropriating economy (when the fruits of nature are used) to the producing economy. In fact, during the preagrarian period, economy had been out of the question in the primitive community of the hunter-gatherer, since humans behaved like all animals, feeding on natural fruits. The producing economy became a method of successive estrangement, taking from nature not its fruits, but the whole natural vegetative and animal mass over specific sites of land and, substituting them for one plant or animal domesticated by humans.

Though hunters had domesticated dogs, domestication of other animals and the transition to stockbreeding had taken place during, or soon after, the appearance of agriculture. Humans domesticated sheep, goats, pigs, and cows. Much later, in the

Iron Age, horses and camels were domesticated. Traces of evidence of breeding domestic animals found in Neolithic Egypt are 7,000 to 8,000 years old, in the Middle East 6,000 to 7,000 years old, and in China and Europe 5,000 years old. Initially, domestic animals had been used as a source of meat, wool, hides, then milk and only later on as pack, carting, and working animals.

The agricultural revolution had become a turning point in humankind's development. Now, one human needed one hectare of land to feed on instead of 500 ha as a gatherer, or 5,000 ha as a hunter, and he could feed not only himself but also his family. Agriculture needed new tools, a means of processing the product, and a capacity to store the yield and seeds. This stopped nomadic life, making humans lead a settled life, giving them time for thinking and other activities. This accelerated cultural specialization; exchange and trade appeared and trading rules were established. Settlements, prototypes of towns, appeared as meeting points for exchange of agricultural and craft products. Special people monitored the rules of exchange to be accomplished, they settled arguments, protected producers – so officials and warriors appeared (i.e., power). Gradually, modern civilization was forming and many of its features originated from deep antiquity. Human's views of the environment changed, too. They began to feel masters of their own environment.

Both gatherers and hunters deified nature. They worshipped the trees and animals that gave them food and the natural phenomena that completely governed their life. After the Neolithic revolution a gradual change in worship and gods took place. In ancient Egypt, gods acquired a human-like appearance, though some of them preserved animal's heads. Gods representing natural forces, however, remained (such as Ra, personifying the Sun) and continued to be worshipped for agricultural purposes. Such a mixture of human and animal features is inherent in the early religions of many agricultural countries. But in ancient Greece and ancient Rome, gods, though personifying various natural forces and elements, became like humans. They behaved on Mount Olympus like humans: they loved and were jealous, envied and tried to outwit, made friends and quarrelled. At the same time, in contrast to humans, they were endowed with a huge additional energy, which made them able to do much more than humans. Such a depiction of gods testified to the fact that the ancient crop grower had become aware of his power and this put him above nature. When he changed over from polytheism to a single god, he endowed him with an omnipotent power. He also imputed to him the following words: "And said the God: let us create a man just like Me, and let him dominate fish in the sea and birds in heavens, and cattle, and the whole land, and all reptiles creeping on the earth."

The appearance of agriculture became a very important event in the man–nature relationship. It initiated civilization, which still remains agricultural (agrarian) despite all the achievements of scientific and technical progress. Agriculture brought forth most of the present attributes of civilization: power, economy, science, religion, etc. One should note that at that time, consumption had been balanced: if a human community had exceeded the admissible level of consumption of the products of natural ecosystems in its territory, part of the community would starve and die, with the remainder moving on to another place. In that way a natural balance had been preserved between admissible consumption and the size of the population. Such an economy prohibited growth but permitted development.

The present economy is appropriating–producing, based on growth. Within this system, humans continuously and increasingly appropriate what should be distributed among other organisms. Meanwhile, more and more energy, in the form of organic matter produced both by cultural and natural plants (forest cutting, fishing, and obtaining other sea products), is transformed into anthropogenic channels.

Before the Neolithic revolution, human beings conquering the planet and mastering new ecosystems (landscapes) had acted as species-violators of ecosystems, like other herbivors. After the Neolithic revolution, a new process had begun, which might be called a new conquering of the planet. It represented the first technological revolution.

First technological revolution

The English archaeologist Gordon Child who called the transition to the appropriating–producing economy "the Neolithic revolution", believed that the transition to food production – deliberate growing of edible plants, especially crops, as well as domestication, breeding and selecting of animals – had been an economic revolution, the greatest in the history of humankind after the mastery of fire. This had opened up the possibilities to resort to a richer and more reliable source of food, which was now controlled by humans, giving them almost boundless possibilities and demanding from them only to apply all their forces.

The transformation from gathering into deliberate cultivating of crops had been a slow process and continued for 2,000 to 3,000 years in the Old World and 3,000 to 4,000 years in the New World. Table 4.1 gives a database mainly on the study of the bone remains of animals, findings of grain repositories, and primitive agricultural

Table 4.1. Chronology of domestication of some plants and animals.

Years ago	Plants	Animals
16,000	–	Dog
12,000	Barley, wheat, rice	Sheep, goat
11,000	Pumpkin, pepper	–
10,000	Wheat, cucumber, plum	–
9,500	–	Pig
8,500	Flax	Horned cattle
7,000	Maize, haricot bean, beans	–
6,000	Vine, millet	–
5,000	Tea, olive	Horse, donkey, cat, bees
4,000	Oats, rye	Hen

tools. Most ancient centres of agriculture discovered in the Middle East appeared 12,000 years ago or a little earlier. The most ancient centres of agricultural civilization appeared in the Middle East and in North Africa (i.e., in places that were home to the ancient populations of humans). About 8,000 years ago, in the period of the postglacial climatic optimum (the Atlantic time), the Sahara desert had a more humid climate, with rivers, and with Lake Chad the size of the Caspian Sea. Here, in the territory of Lybia, were found (in the Lungal oasis) silicon sickles, reaping knives, and grain grinders dating back 8,000 years (Bondarev, 1998). Another centre of agriculture formed in China, in the valley of the Yellow River, and then – in India, in the valley of the Indus River. At present, data appears about ancient centres of agriculture in north-western Thailand and Indochina, but this still needs verification.

From centres of agriculture, civilization gradually began to propagate around other territories.

4.3   GLOBALIZATION OF AGRARIAN TECHNOLOGIES

After the Neolithic revolution a new wave of globalization started, connected to the agricultural technologies, but at a higher level. It propagated more rapidly from centres of cultivation and stockbreeding. From the Near East, agrarian civilization moved eastward – to the Iran upland and Central Asia. The cultivation of cereals started here between 9,000 and 10,000 years ago. In southern Turkmenistan, the Neolithic culture of settled crop growers and stockbreeders appeared 8,000 years ago. In Iran, both cultivation and stockbreeding developed, mining for copper deposits also developed and, hence, metal agricultural tools appeared (Bondarev, 1998). Over the whole area from the Nile valley in North Africa to Central Asia, agriculture mastered new technologies, such as irrigated cultivation. It is of interest that these technologies of agrarian production appeared in the first centres of cultivation. Some 6,000 years ago, in the Nile valley basin, irrigation was practised. During the floods, the Nile water smoothly and gradually filled the flood plain, having broken its banks, and covered the soil surface with a layer of water 0.5–3 m lasting 6–8 weeks. The floodwaters deposited fertile silt containing up to 2% organics. During the flood, water washed out salt preventing the soil from salting up. Moreover, pictures of ancient Egypt show that during the harvest only the spikes of plants were cut, while stems (stubble) were left in the field. This stubble prevented the soil from eroding, favoured silt deposition and served as a fertilizer. Some 4,000 years ago, the Egyptians invented the shaduf – a device to raise water from a well – and 1,000 years later they invented the water-lifting wheel. In Mesopotamia, in the lower Tigris and Euphrates – in a boggy region in the period of the El-Ubeyd (6,000 to 7,000 years ago) – human beings learned to drain soils by bog reclamation, irrigation, and the building of dikes. The inconsistent regime of floods on the Tigris and Euphrates did not allow the use of an irrigation system as in the Nile valley; therefore, in these areas water was sent to the fields with the help of shadufs and, during flooding, through sluice gates the dike water was drained into the canals (Losev, 1989).

In another centre of agricultural civilization – north-western India – the settled crop growers and stockbreeders appeared about 9,000 years ago. On the basis of the first communities of crop growers and stockbreeders, 4,000 to 5,000 years ago in this region a civilization called Harappa reached a high level of development. It was called Harappa after the name of the village near the ruins of a large settlement discovered in the 1920s. This civilization was in no way inferior to civilizations in Mesopotamia and Egypt. The Harappa dwellers grew wheat, rice, vine, date palm, and sugar cane. They were the first to cultivate cotton plants, and to learn to spin and weave. Two large towns have been well studied – Harappa in Punjab and Mohenjo-daro in the lower Indus (here irrigated cultivation was also practiced). From here the cultivation culture propagated over the entire Indian subcontinent.

The very eastern centre of the agricultural civilization in China on the Yellow River gradually moved southward to the basin of the Yangtze River and farther eastward – to Korea and Japan. In Japan the agricultural technologies were mastered only at the closure of the Dzemon culture (about 3,000 years ago). At that time in Japan they began to grow dry land rice, millet, and buckwheat. The eastward movement of the agricultural civilization from the Near East was towards the Balkan Peninsula, where, as indicated by archaeological data, the centres of cultivation appeared 6,000 to 8,000 years ago. Later on, cultivation appeared on the Apennine Peninsula. By this time (i.e., 5,000 years ago) the wooden plough was invented to till loose soils on alluvial river terraces. Gradually, the centres of cultivation covered all the known parts of the ancient Old World, where climatic and soil conditions favoured agriculture. So, archaeological data testify to a penetration of the carriers of the culture of boat-shaped axes already familiar with primitive cultivation and stockbreeding, to the outskirts of Oikumene and to the Baltic countries about 4,000 years ago. Some 1,000 years later, stockbreeding remained the main source of existence in these areas, although cultivation continued to develop. Only at the beginning of a new era had cultivation become the basis of the economy.

In America, cultivation appeared independently, but later than in the Old World. Here a civilization of tropical cultivation developed. The precursors of the Maya in the south-eastern part of Mexico took up cultivation 3,500 to 4,000 years ago in the mountain regions of the Pacific coastline. About 3,000 years ago they developed a forest plain in the south of the Yukatan Peninsula. Here dozens of towns were built. Mayans were not familiar with the wheel, plough, draught animals, metal tools, or the potter's wheel, and still lived, in fact, in the Stone Age. They grew maize, beans, pumpkin, manioc, and the bread tree. In the New World, the Peru–Bolivia Plateau (Altiplano) witnessed the first centres of cultivation. It is here in the Peru–Ecuador–Bolivia centre (selected by Vavilov, 1926) that the origin of many cultivated plants is located, including maize, potato, tomato, pumpkin, and tobacco. Cultivation began here 8,000 to 9,000 years ago. In caves, traces of pumpkin and pepper cultivation have been found. Some 10,000 years later they started growing maize here – the main grain crop in America. The earliest use of potato, dated 5,500 years ago, was in the basin of Lake Titicaca. Cultivation also developed along the Peruvian coastline, where it was combined with fishing. However, within the vast territory of America, only small enclaves of agricultural civilization appeared. These propagated slowly. Besides, there were catastrophes eliminating whole regions of agricultural production – 900 BC, a tsunami, more than 40 m high hit the Peru shoreline. For reasons still unknown, in the ninth century the agricultural civilization of the Maya collapsed, and their settlements become overgrown with tropical forests. Apparently, a connection between the South-American and Central-American regions of cultivation did exist, since potato and maize got to Central America from their first land – South America (Bondarev, 1998).

Nevertheless, agricultural civilizations with all their repeated crises in many territories, continued to propagate. After the disintegration of the Roman Empire agriculture fell into decay in Europe. This period is known as the Dark Ages and refers to the beginning of our era and the earlier Middle Ages. At that time the Roman economy degraded. The agricultural fields were shrinking, water systems went to ruin, the towns built by the Romans became depopulated, many roads built by the Romans became overgrown with grass, and neglected fields were again overgrown with forests.

Only from the 9th century did agriculture rapidly begin to develop in Europe. The King and later the Emperor Karl the Great (742–814) issued an edict which granted to everyone a site of forest under the condition that it would be deforested and stubbed up to obtain an agricultural field. Development of new lands – virgin forest soils – began in the 9–12th centuries (most actively in Europe). It was called "the great uprooting" (Bondarev, 1999). Forests were cut for agriculture, construction, charcoal, and to build ships. By the beginning of the 20th century, western and central Europe, earlier covered with forests, became a territory of almost continuous agricultural fields (Figure 4.2) (Dorset, 1968).

At the same time, the regions of first agricultural centres were almost completely developed. The subcontinent Hindustan, Central Asia, the eastern part of China, and South-East Asia became territories of agricultural production. On the whole, it took almost 10,000 years for the Old World to cultivate the basic, most suitable, and even unsuitable lands.

In 1492 Christopher Colombo discovered America, and a stream of emigrants rushed from the Old World – the carriers of agricultural civilization. The newcomers started to rapidly populate new lands, developing agricultural production and the cultural plants of the New World. In North America the agricultural technologies began their triumphant movement westward. First, the western part of the USA was developed, to the Mississippi River – the Great Plains. By 1830 all these territories had been transformed into agricultural fields and pastures.

Agricultural development of the US territories

In the USA (which is similar to Europe in size), agricultural development of the territory took 200 years to effect, whereas in Europe it took almost 10,000 years. Parson (1969) characterized the development of the US territory in the following way: "the westward movement of the colonists was followed by an equally huge extermination of natural resources. In the west, the boundaries were broadened with an axe, musket, and plough". By

1900, the USA had become the greatest agricultural power, and during the
First World War, having ploughed new millions of hectares of soils,
provided Europe with food.

Australia was also rapidly populated, with conditions suitable for agricultural
development. Agrarian technology propagated all over the globe – agricultural
globalization had taken place.

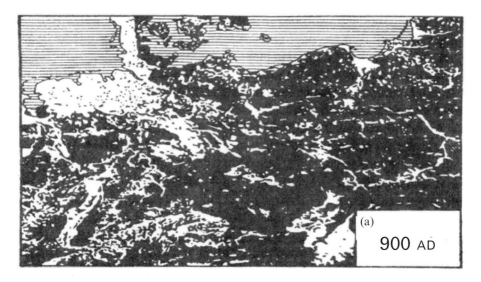

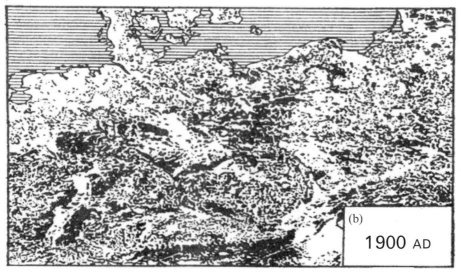

Figure 4.2 Extent of forestation (dark areas) in central Europe for the years 900 (a) and
1900 (b).

As was mentioned above, the Neolithic revolution had changed man–nature relationships, and had also changed man himself. Instead of a balanced relationship there appeared an unbalanced one, resulting from the destruction of natural ecosystems and their subsequent replacement by artificial environments designed to provide for human needs and comforts. Without a moment's hesitation, humankind had destroyed vegetation and animals in the interests of its own. The last aurochs died in Europe in 1627, and as far back as the 9th century, Karl the Great hunted aurochs in Aix-la-Chappelle. In the USA, from 1879 to 1875, up to 2.5 million bisons were eliminated annually. Consequently, after 1889, bison became an endangered animal.

The entire history of agricultural civilization represents destruction of nature followed, as a rule, by resource and food crises, leading to a decline in the population. The usual scheme of such crises was a chain of ecologogical–social–economic events: development of new agricultural lands, increase of food production and growth of population, therefore tax collection. This was followed by reduced soil fertility due to erosion, a lower level of agriculture, increased taxes (since the authorities tried to preserve the budget level), impoverished peasants, hunger, and weakening of the state. All this led to rebellions and uprisings, often with the conquest of the territory by newcomers. Such chains of events had repeatedly taken place, often followed by natural cataclysms. For instance, several low-yield years and hunger were the prologue to the French Revolution.

Over the 10,000 years of agricultural civilization, natural ecosystems had been severely damaged, especially those most productive ecosystems – forests. Both gatherers and hunters had burnt forests, clearing space to attract large herbivorous animals, to ensure their own safety. This sometimes led to unintentional fires. The Neolithic revolution had brought to life a slash-and-burn form of cultivation, widely used in various regions of developed agriculture – from the tropics to moderate latitudes and even wooded tundra. Various estimates, from the beginning of humankind's violation of the forest ecosystems to present-day suggest that up to half to two-thirds of forest areas have been eliminated. Table 4.2 shows that during the time of this agricultural civilization between 500 and 750 Gt of carbon were emitted to the atmosphere (due to elimination of 30–40% of forests) while the natural production of forests was reduced by 26–40 Gt of carbon per year.

This did not mean that the carbon released in the deforested territory had been

Table 4.2. Produce, productivity, and biomass of the present forests, grass, bush, and arable land (Gorshkov, 1995).

Territory	Biomass (Gt C*)	Produce (GtC yr^{-1})	Productivity (tC ha yr^{-1})	Area (10^6 km^2)
Forests and marshes	750	40	8	50
Grass and bush	50	13	4	33
Arable land	5	5	3	15

* C – carbon

totally substituted by that of arable lands or pasture (grass and bushes), since deforestation and retreat of forest boundaries to higher latitudes had led to a decrease in transpiration and, hence, in the rate of the continental water cycle, which in the northern hemisphere favoured the desertification and the northward retreat of the southern boundaries of forests and steppes.

Reduction of the continental moisture exchange

As is known, to produce 1 g of organic substance, a plant needs 100–1,000 g of water to transmit through its leaves. Some 1,000–10,000 g of water are spent to obtain 1 g of organic carbon. Hence, a decrease in productivity of the forest ecosystems leads to a decrease of transpiration. The transpiration capacity constitutes two-thirds of the capacity of total evaporation (see Table 1.1). Leaves of the forests absorb and utilize (on transpiration) 90% of the absorbed solar radiation, whereas in anthropogenic agrosystems only 40% is utilized (i.e., agrosystems almost halve the evaporation compared to forest ecosystems). On the whole, the estimates show that the value of transpiration during the period of agricultural civilization reduced by tens of thousands of km^3 (i.e., almost by the value represented by present-day river run-off).

Desertification due to deforestation was observed in many regions of the globe, including the centres of ancient cultivation – in the Near East and in the valley of the Indus River. In the eastern region of China where the monsoon climate prevails, it was less evident. But in the central part of China, in the west and in the northwest there are vast deserts the size of which increased, of course, after deforestation in the eastern part of the country. The Russian soil scientist Kovda (1990) calculated that at least 20 million km^2 of soils previously used as arable lands and pastures, were then lost due to desertification. Thus, only a smaller part of deforested territories became arable lands and pastures. A considerable part of these lands later turned into deserts. Hence, a reduction of the global ecosystem's produce due to desertification was only partially compensated for by the produce of arable lands and pastures. Usually the pastures' produce (grass and bush) is less than that of arable land, which in turn is less than the forest's produce.

These data show that the agricultural civilization strongly affected the biosphere of the planet. The continental moisture exchange was sharply reduced and the land-surface albedo changed. The erosion of agricultural lands resulted in a growing dust content within the atmosphere and river run-off of alluviums. Due to deforestation, the atmosphere accumulated a huge amount of carbon dioxide, with soils and water objects accumulating large amounts of nitrogen and phosphorus compounds. The development of the planetary system of agriculture was exponential and the greatest associated changes in the environment fell within the last two centuries of the second millennium.

4.4   SECOND TECHNOLOGICAL (INDUSTRIAL) REVOLUTION

"The great uprooting" had led to an economic boom in Europe: towns grew, crafts developed, first universities were founded, etc. Watermills were widespread, the manufacture of glass and paper began, windmills were built, gunpowder appeared, as well as the compass, glasses, steering wheel, cannons, and, finally, book printing was invented with the use of engraved wooden boards and later block letters of metal (Bernal, 1956). All this was realized in the period of the so-called developed Middle Ages (11–14th centuries) and resulted from rapidly broadened areas under crops and pastures, improved agrotechnics (two-field system was replaced with the three-field one), an appearance of a wheeled plough, horse collar, horseshoes, and the use of fertilizers (Bondarev, 1999). In the mountains of central Europe, thousands of mines were developed. Melting of metals, salt-works, and smithies needed great amounts of wood and charcoal, which led to rapid deforestation around these works. Trade and banking increasing, as well as usury, which opened up new sources of enrichment not connected with power or land property. Science and arts were developing. Enlightened people took an interest in the antique and Arabian sources. So the epoch of Renaissance began, which put humankind at the centre of the world, forming an anticlerical and antifeudal direction. Economists and sociologists called the period from the beginning of the Renaissance to nowadays the "modern" period.

At the same time, three parallel processes took place in Europe. Large flows of money gradually formed which created the basis for future investments. They were formed via the legal robbery of people – through increased length of working days, raised rent for poorer tenants (serfdom had already been abolished), expropriated community domains, and raised taxes. Financial gain also increased due to formation of trade capital growing on the price difference in areas of production and sale, creation of the usury and of banking capital, as well as by robbery of colonies. Finally, capital was also accumulated by means of piracy.

The second process is the development of science and creation of new technologies. The process advanced due to broadened education, book printing, the appearance of the first scientific journals, a growing number of universities, an improved way of waging war, and development of manufacturing.

Finally, the gradual process of liberalization of society took place, first by religious liberalization – independence of religion on a national basis – as well as by the appearance of trade republics (Venice and Genoa) and constitutional monarchies (the Netherlands and Great Britain). In the 17–18th centuries these three processes were supported by the ideas of the Renaissance. The formation of a liberal economy (free market) and liberal civil society took place, but a long way off future historical development lied before them. The Netherlands and England were the first countries to walk this path, introducing liberal governmental structures.

In England, the creation of capital flows began after the development of smooth woollen cloth mills, where production required large quantities of wool. To broaden the pastures, landlords ousted the tenants from land by raising the rent, expropriating community domains, draining the bogs, and cutting the forests. Homeless and poverty-stricken peasants became a reserve of cheap labour. At the same time, large trade companies and banks were organized in England, exploitation of colonies began, and piracy was encouraged.

Deforestation as a stimulus of development of new technologies

An important stimulus to develop new technologies was a liquidation of forest ecosystems in Great Britain. As a result, a power crisis took place, which led to the use of coal and a shortage of timber as a building material. This stimulated a change of technology in metallurgy, a creation of Cartwright's weaving loom and Watt's steam engine, as well as raising the efficiency of production and labour productivity, based on fossil fuel energy, steam engines, and machine tools. All this was used to organize the mass production of goods for the population and marked the beginning of indus-trialization. Finally, the constitutional monarchy in England guaranteed an inviolability of private property.

A structural reorganization of production took place, which liquidated craft workshops and shop systems of production and created large factories, plants, and industrial centres. For the first time, huge masses of wage labourers appeared. This phenomenon was called the "industrial revolution", but, in fact, it was the beginning of the formation in the world of liberal market economics and civil society. First, it was believed that plant and factory workers ought to be provided with food; second, based on the development of industry, new agricultural machinery was developed (row ploughs, horse draught harrows, cultivators, reapers, etc.), and in Europe new cultured plants were introduced from the New World (potato, tomato, maize, etc.). Thus, the industrial progress – the second technological revolution after the Neolithic – determined the first "green revolution" in agriculture.

Development of the territory of the present USA occurred mainly at the same time as the Industrial Revolution. Then, from Europe came immigrants who brought agricultural knowledge and industrial technologies. Though in the 19th century the USA remained an agrarian country, the rate of industrial development was high. In 1800, the rural population constituted 96%, and in 1850 it was 87.5%. In 1840, the textile products from the USA successfully competed with those from Europe. A stimulus for industry was the development of agriculture using machines designed and manufactured in the USA. In 1851, at the London World Exhibition, the USA exhibited most of the agricultural machinery. In 1807, R. Fulton was the first to build the wheeled ship "Clermont" which was exhibited on the Hudson River. In the 19th century the USA topped the list of railway construction, and in 1894 led the field in industrial production, surpassing England, the first industrial power.

The industrial revolution marked the beginning of industrialization in many countries of the world, initially Europe. This revolution was not only a technological and economic process but also a social reorganization, a transformation of a traditional rural, patriarchal, holistic society into an industrial, urban, democratic, and individualistic one.

The process of accelerated modernization of society started in the 19th century from industrialization, continuing until the end of the 20th century, with some countries still undergoing the process. But by the mid-20th century, modernization became a global phenomenon. It should be borne in mind that it took thousands of years for globalization of the agricultural society. *Now the flows of capital aimed at industrialization and obtained by robbery of the people began to work for continuous economic growth at the cost of an accelerated robbery of nature.*

In the process of modernization, three features should be noted. At the beginning of the last century a new social–economic system was developed in Russia – the centralized management of the economy and a totalitarian regime. But this system, while trying to preserve the attributes of the holistic agrarian society, also developed industry using the same instruments as the market system. The totalitarian regime also created the financial flows generated from the robbery of the people and then started to reorganize nature. The planned system, like the market one, was aimed at economic growth. In fact, it was the "dissident" form of modernization.

The second point to note was that modernization seriously changed agricultural production, putting it on industrial rails. Energy consumption greatly increased, machinery was modernized, synthetic fertilizers appeared, protection of plants intensified, and a new variety of plants appeared as well as new breeds of animals and birds. All this made it possible to, more or less, meet the growing need of food through increased efficiency of agricultural production. This process was called the "green revolution" and this was the second "green revolution" in agriculture.

Finally, modernization accelerated an active growth of towns and urban populations. The process of urbanization rapidly individualized people, developed their self-conscience and directed them towards a search for ways of realization of their abilities.

The industrial revolution and industrialization created a greater freedom and a wide range of possibilities in mass consumption of natural resources. The rapidly developing science provided new technologies to realize even more acute ways for expansion of production and use of natural resources. Industrialization, first of all, required a space to build the manufacturing complexes and to extract raw materials. Also, it was necessary to create a developed infrastructure including, first of all, roads. All this accelerated the destruction of ecosystems, putting in their place towns, roads, and airports.

Industrialization also formed a brand new source for affecting the environment – physico-chemical pollution. Industrial production and infrastructure are based on chemical reactions, physical phenomena, and processes. *Though many of these processes repeated natural processes, human beings can not create a closed cycle of a substance in developed industries.* Mineral and organic raw materials, grown in the fields and pastures as well as extracted from natural ecosystems, are transported and processed to achieve a final product. At all stages of this process, waste appears, and the mass of the final product (without an account of raw materials) constitutes only several % of the raw materials. Everything that remains is waste. With an account of raw materials the percentage of its use increases. The final product for human consumption sooner or later also becomes a waste. Calculations show that 1 kg of discarded garbage is equivalent to 25 kg of production waste. Thus the entire industrial power of the world is, in fact, a mechanism of waste production. Consumer goods can be short-lived, becoming waste within 1 year (some articles of domestic use), mid-lived, becoming waste between 1 and 10 years (devices of everyday use, cars, etc.), and long-lived, becoming waste after 10 years (buildings, roads, etc.).

The basic mass of waste is solid substance, and only 10% are gaseous waste and dust entering the atmosphere. Substances together with solutions or suspensions get into water bodies. To process solid waste, space is needed, providing one more means to liquidate natural ecosystems. Air and water pollutants are spread by air and water fluxes affecting both natural biota and humans, accumulating in soils, plants, bottom deposits, and in living organisms (the human body included).

The world resource-consuming economic system

The world economic system is very wasteful. For instance, only 3% of the energy of an atomic or coal power station is transformed into the luminous energy of an electric bulb. Only 15–20% of the energy in car fuel is spent turning the wheels of a car. In the USA, 80% of products are used only once and most of the remaining products not more than twice. Finally, 99% of the initial material spent on a given product and contained within it, become, on average, waste in 6 weeks after sale. The Americans pay up to one billion dollars per year for real but unused energy and other materials, with, on a global scale, 10 billion dollars being spent per year (Weitzsäcker et al., 1997).

Thus the second technological revolution became a new powerful attack on the biosphere. It also changed people's lives, transforming humans into urban citizens, in many cases providing a countryman with the attributes of urban life. It individualized human beings and reduced their close relationships with other people and distant relatives. It transformed humankind into an experimental rabbit, unwittingly testing the effects of 18 million chemicals as well as the impacts of many physical phenomena (from light and sound to various types of electromagnetic radiation). But the main "achievement" of the second technological revolution based on the ideas of modernization was an undermining of natural mechanisms of regulation and stabilization of the environment – ecosystems (landscapes) destroyed or severely damaged by human activity over 63% of inhabited land (15% of the Earth's surface), which has led to a global ecological crisis.

4.5   THE THIRD TECHNOLOGICAL REVOLUTION – THE INFORMATION AGE

The end of the 20th century was characterized by an unprecedented rate of propagation of information technologies, initially in developed countries. Information technologies are a means of rapid processing and transmission of information. Developed countries have moved on from extensive to intensive development. Transition to the *resource-intensive development* meant that, from a smaller mass, a greater number of product units are obtained, with smaller expenditures of energy and water. A reduction of water expenditure per product unit is an important indicator, since, by mass, it is the raw material most frequently used in the world – at least an order of magnitude greater than the mass of other raw materials. These cardinal changes ought to have helped the world community to transform the economy, to solve the social problems accumulated from the beginning of industrialization, and finally, to solve ecological problems. However, this did not happen.

Nevertheless, many economists, sociologists and politicians write and speak about the transition of many countries from the industrial society to the information society, which can potentially solve many global problems. However, facts contradict these expectations. No matter how the economists and politicians call the future world, it will remain both agrarian and industrial. Food, metal, fossil fuel, plastics, and chemical materials will be always required, meaning that both primary and secondary economies will exist and develop. Reducing the traditional industrial production in the USA and Western Europe is contradicted by their transfer, to the developing countries, of such resource-consuming and "dirty" technologies producing semifinished products or details (rails, tubes, steel sheet, etc.) to produce final products in the developed countries. Therefore, in the USA a reduction of metal melting did not lead to a decrease in car production or military machinery production. On the contrary, it increased. The same is observed in Western Europe. Closing the coal mines in Germany did not reduce the growth of energy consumption because coal was replaced by gas exported from Russia. In the near future the world can reject neither industry and industrialization, nor agriculture. Examples of the East Asian "tigers" (South Korea, Taiwan, and Singapore) and China show that industrialization is very much ongoing. One can only state that an international division of labour has taken place, when more and more "dirty" material, energy, and water-consuming production is transferred to the developing countries. This has occurred, first of all, for the following economic reasons: closely located raw materials, cheap labour power, absence of serious ecological legislation, and creation of infrastructures for transferred developed country enterprises, often at the expense of the host government. Such a "manufacturing" globalization has led to a stable division of the world into the main developed countries, a group of countries attending to them, and the poorest countries, which have practically fallen out of the global economy system (Losev et al., 2001b).

An intensification of economies in the developed countries which began in the 1970s led neither to a rapid growth of labour productivity nor raised the rate of the gross domestic product (GDP). Moreover, the rate of the GDP growth in the

1980–1990s was reduced by more than one-third compared to the period 1965–1973. During this period the global volume of investments was low (developed countries included). During the last decade of the 20th century the volume of investments into new enterprises as well as into equipment and new technologies decreased. The debts of the developing countries to the developed ones and their share with respect to gross national product (GNP) increased. Capital rushed from the developing countries. Before 1983 the financial flow was directed to the developing countries. Presently it flows back to the developed countries, in 1985 it exceeded 20 billion dollars annually (Danilov-Danilyan and Losev, 2000).

Usually, both informatization and information are supposed to be stimulating society by increasing gross domestic product (GDP) and people's well-being. Meanwhile, the US policy makers observe the following paradox: most people feel that while the GDP grows markedly, "they have to run faster in order to remain where they are or even to lag behind a little". Incidentally, this is not only "felt by people", but demonstrated by the statistical data of the developed countries. In 46 of the most successful countries the standard of life of 19–40% of the people reduces every year. In the developed countries there are 35 million unemployed. In the USA – one of the richest countries (32% of the profit of all developed countries) – from 1975 to 2000, 1% of the richest people increased their share in the finances of the country from 20 to 36%. Nevertheless, the income of an ordinary American family did not increase despite the fact that the average number of working members of the family increased (mainly women). The amount of poor (by US standards) during the period 1959–1977 decreased from 40 to 25 million people, but by the mid-1990s it increased to 36.4 million people (Danilov-Danilyan and Losev, 2000).

An increase of GDP, as many economists and ecologists believe, does not reflect a growing well-being of the population of the country. Two organizations – New Economic Foundation together with the Stockholm Institute of the Environment – developed the index of sustainable economic well-being (ISEW), which was also called the genuine indicator of progress (GIP). For the USA, Great Britain, and other developed countries, the calculation of this index has shown that in the second-half of the 20th century a correlation between this index and GDP ceased in the early 1970s. The index value dropped sharply, but the GDP kept on growing (Figure 4.3).

In fact, neither GDP nor GNP reflect real life or people's well-being. Figure 4.4 shows that with the growth of GNP, both the capital gain and the index of the growth of real income of working people, starting from the 1980s, has made no headway. This testifies to the development in the world of a social crisis (developed countries included).

Formation of GDP at the expense of social and ecological costs

Let us take the example of a collision between two cars. As a result, a complicated mechanism is triggered (ambulance, treatment, insurance, hearings, reports in newspapers, etc.), which, from the economic point of view, are considered as professional activities that need to be paid for. The

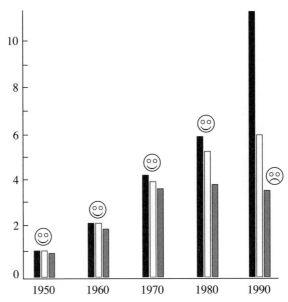

Figure 4.3 Change in gross domestic product (GDP) in US dollars (white columns), net income from capital (black columns), and in the index of sustainable economic well-being (ISEW) (grey columns) in the developed countries in the second half of the 20th century.

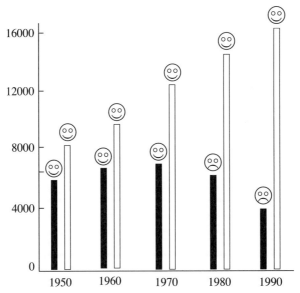

Figure 4.4 Change in the pure capital gain (black columns), the gross national product (GNP) (white columns), and of pure real income of working people (black columns) in the second half of the 20th century.

victim of the accident bears losses but the GDP grows. Here is another example. Although for a baby mother's milk is the best possible food, suckling a baby reduces GDP. Artificial feeding raises GDP, since in this case there is a need for bottles, cow's milk, teats – products of large enterprises. Similarly, environmental degradation is favourable for GDP growth, since it causes economic activity by creating the need for infrastructure to carry out nature protection technologies. Although environmental violation has a deleterious effect on people's health, the infrastructure for nature protection technologies is arguably one more mechanism for ecosystem destruction, pollution, and resource consumption (Weitzsäcker et al., 1997). Finally, the extraction of resources contributes significantly to GDP, but continued extraction itself leads to an exhaustion of resources and reduction of possibilities for future generations to ensure their well-being as well as ecosystem destruction and environmental pollution.

National GDP accounts provide information about employment and are used to assess income taxes, which was first the priority of policy for a long time, as specialists noted. But, it cannot reflect the whole spectrum of social well-being.

In the 1980s, as a result of the broadened free international trade, the capital gain started growing rapidly, and the income from real material production decreased (Figure 4.4). Now, fluctuations in the financial market have begun to increasingly depend on its own financial laws but not on material production, material flows, and economic decisions. With an increasing internationalization and disparity between real and monetary spheres, the probability of the world economic crisis and worsening of the socioeconomic situation grows most noticeably in the developing countries.

Local financial crises took place in the countries of South-East Asia, Japan, South and Central America, and Russia. The first sign of the crises were noted in the USA and Europe in April 2000, when the value of shares in the largest firms in the spheres of high technologies dropped in stock exchanges. At the end of 2001 a financial crisis hit Argentina, and in 2002 it began to develop in Uruguay.

Exaggerated economic assessments of the sector of high-tech technologies in the USA

In the 1990s the price of shares in the so-called "high-tech", science-based technologies rapidly grew in the USA. American society was drawn into financial intrigue: over 10 years the number of families owning shares almost doubled – from 28% in 1989 to 54% in 1999. The capitalization of high-tech companies greatly exceeded the size of the annual profit (i.e., their price was overestimated). The economic recession which started in the USA revealed the real state of affairs in the tertiary economy of the country, and in 2002 the price of shares in such companies dropped.

New rules of competition have been established in the financial markets that go beyond the realm of ethics. In this market, hypercompetition has transformed into financial war with the objective of ruining competitors. Such immoral competition is inherent not only in speculative international financial markets but also in the global movement of finances. It has even spread to the material sphere of production (Weber, 1999).

Thus substantial changes are observed in the global economy which neither improve the socioeconomic situation in the world nor stabilize the economy despite a GDP growth. Owing to information technologies, finances can be transferred to any place on the planet (for 20 cents, one million dollars can be transferred in one minute). The high speculative capital gain is fraught with financial crises, like the 1998 crisis in several countries. Information technologies and networks (banking, the Internet, etc.) have opened up new possibilities, especially for financial flows and trade. This branch of economy developed very rapidly. The number of Internet users increased from 3 million in 1993 to about 200 million in 2000, which exceeded by far the growth in the number of radio and TV users. Even in Russia there are places where individuals can at any time, through a computer, take part in a game at the international currency markets of New York, London, and Tokyo.

The reduction in computer prices was unprecedented (30–40% annually), never happening with any other product. The world had hopes for computerization and information technologies, especially in the sphere of the economy. There were hopes to increase the rate of economic growth (GDP growth) and labour productivity. Earlier, such hopes had been raised by the scientific–technical information (STI) computerized systems developed in the 1970s. At the 1978 Atlanta Conference its participants – representatives of five countries with the most developed STI systems at that time (the USA, the UK, Japan, France, and the USSR) – discussed only one problem: why the developed STI systems, with a fast access to various sources of information, have not accelerated the scientific–technical progress and have not raised GDP and labour productivity. This question could not be answered then, nor can it today. Probably, the development of such a system serving science and engineering was a necessary step and facilitated the work in this sphere, but could not accelerate the appearance of new ideas of scientists and engineers.

With the development of new information technologies, communications and global networks, the same question arises, still with no answer. The global propagation of information technologies over the past 20 years neither increased GDP nor labour productivity. The Nobel prize-winner in economics Robert Solow characterized this phenomenon in the following way: "Indications are seen everywhere to the onset of the computer epoch, except statistical estimates of labour productivity." This phenomenon is called "the Solow paradox".

Computers do not generate new ideas

At present, computers are widely used (e.g., to calculate and predict river run-off). The mass intrusion of geoinformation technologies into this sphere has taken place. However, on the whole, the accuracy of calculations and predictions has remained static. In this sphere, as was mentioned at one international conference, progress is slow despite a variety of models and an abundance of information technologies. This reasoning is explained as the incomplete understanding of real processes, a great deficit of data, and the absence of methods to determine important parameters in the models. Computer and geoinformation systems cannot obtain concrete data and give new ideas, but they can rapidly process the data, display it, and obtain results with the use of the most complicated models consisting of hundreds of equations, however, it is still necessary to estimate the degrees uncertainty.

Most likely, the development of information technologies will be one more step toward STI progress. Informatization has changed our world, introducing many positive effects, but also damaging ones. So far, it is difficult to completely assess this damage, but some of them are apparent even now. Computer games have become a mass enthusiasm for both youngsters and adults to the detriment of other spheres of cultural heritage – science, the arts, and sport. Most of the games include elements of violence, induce an unhealthy excitement in the game player, and add to noise pollution. Information flows within and between all the computers of the world constitutes about 10^{16} bit s^{-1}. This exceeds (by 6 orders of magnitude) the natural ability of humankind (constituting not more than 10^{10} bit s^{-1}) to assimilate information. Therefore, all information above this value contained in computers and in all other sources (TV, radio, video, cinema, etc.) represents information pollution, as ecologists say, or information noise, as specialists in the field of informatics call it. This excess information, in their opinion, slows down decision making and reduces the growth of productivity and efficiency. Hackers who plant viruses in information systems also create information noise, bringing about serious financial losses. From the viewpoint of ecologists, a huge information flow (its excessiveness) affects all our senses, threatening humankind's mental health. Only time will tell what will become of the present young generation, who are becoming more and more involved in the systems of information technologies.

Computer technology is another sphere of industry that violates the environment. For example, to manufacture one computer, 20 t of water are needed during production. Computer equipment is rapidly changing, as it is a product with a short lifetime. For example, in the USA one million personal computers per year are disposed of, with both software and computers being replaced every two years. Something needs to be done with obsolete computers and diskettes. In Europe, each year 6 Mt of waste appears in the form of household appliances that have outlived their usefulness and/or electronics that contain dangerous substances and heavy metals. Since 2002, dissassembly plants have been built for such appliances. It is assumed that 80% of the components can be reused, and from 2008 the use of heavy metals and poisonous synthetic substances within them will be prohibited (*Science and Life*, 2002, No. 1).

Hopes of controlling the biotechnical system with the help of information technologies to provide environmental stability are illusory, since, from available estimates, the information about biota is now 20 orders of magnitude greater than the information flow in all modern computers (see Section 2.1 and Figure 7.1). And this information hole will be hard to fill (Gorshkov et al., 2002).

Thus the appearance and rapid development of information technologies have not made people happier (often, the opposite is the case) and have not better provided people with food, clothes, dwellings, drinking water, or an ecologically pure environment. Information technologies only accelerated the entry of easily obtained information about the growing deficit of all this in the world, into the geographical environment where people live. All technological revolutions turned out to be only a new acute means of extracting natural resources, via transformation into a greater variety of goods, increasing present and shelved waste, and annexing new territories for storage of such goods (Losev et al., 2001a).

5

Formation of the world community and its ideologies

5.1 WORLD POPULATION: FROM GATHERER TO MODERN MAN

Human hunter-gatherers, like other large, moving animals, inevitably destroy biomass over their territory. In the genome of moving animals, strict restrictions on their population are recorded. Studies carried out in the temperate forest ecosystems have shown that annual fluctuations in the size of populations of large herbivorous animals average 2–3 kg of the mass of the animals (boar, deer, etc.) per 1 ha of forest habitat. The same programme of restricted population density is recorded in the human genome. Both gatherer and hunter had been minor violators of the environment, like other animals akin to them.

Studies of the dynamics of the size of populations of moving animals revealed two strategies regarding changing the size of a specimen's population: the so-called K- and r-strategies. Species of K-strategy are characterized by large dimensions, low productivity, long lifetimes, low population densities, and prevailing adult specimens. Species of r-strategy are characterized by small dimensions, high productivity, short lifetimes, variable population size and density, with young specimens prevailing. Large herbivorous animals of the K-strategy include, deer, bison, orangutan, while the r-strategy includes species like the ordinary squirrel and mole.

K-strategy of humans

A human being, being a large herbivorous gatherer, falls into the category of animals exhibiting K-strategy, most notably in the precultivation epoch. One can even state that human beings exhibit an extreme expression of K-strategy, since their lifetimes can be four times longer than animals of the same size. The strategy refers to large animals, with a low birth-rate. Although humans could hypothetically give birth to 30 children per lifetime, the actual birth-rate does not naturally exceed 16–18 children (for an Australian aboriginal). Now, even in the countries with the highest birth-rate, up to 50 children per 1,000 people, and in the developing countries with a low birth-rate – only 10–15 children, the birth-rate falls below what biological productivity permits for humans (200–250 descendants per year per 1,000 people) (Severtsov, 1992).

Available estimates show that the population size of a primitive human con-stituted about 200,000. This slowly grew as humans improved their abilities to obtain food: animal corpses, insect larvae, mollusks, small animals, as well as developing new territories. About 90% of the food ration was vegetable-based.

The human body is almost void of hair cover, which suggests tropical and subtropical latitudes as a place of origin for *Homo sapiens*. Humans were in a state of constant overheating, since they exerted much more energy on their movements over the fodder territory than animals of similar dimensions. They could cover up to 40 km per day (i.e., twice as much as other mammals) being in a constant state of energy deficit, as a result of leading a nomadic life in search of food.

For a human of medium dimensions the basic metabolic power (at a fixed temperature) in a state of rest some time after taking food constitutes 80 W, at a leisurely stroll the metabolic power is about 140 W, and while running it can reach 10 kW (Gorshkov, 1995). Thus there were ecological and biological limitations both to human habitat and growth of the size of their population.

Only after having mastered fire and having learned to make tools and develop technologies (the use of fire, protection from bad weather, storage and transporta-tion of food and fire, use of natural and built shelters, and the use of hunting and fishing tools) could a human go beyond its biologically and ecologically determined habitats, using cultural heritage passed from generation to generation in an extra-genetic way (see Section 1.4). This sharply broadened the territory of his propaga-tion. The ability to generate extragenetic information, improving and creating new technologies made it possible for human beings to develop in the most diverse landscapes – from tropical to high latitudes.

A sharp broadening of human habitats to planetary scales ensured the growth of the population. Assuming the total natural habitat of humans, taking into account the area needed to feed both a gatherer and a hunter, was equal to the present area of arable lands, the size of the population would not have exceeded 2 million people. Demographers, based on other data (the number of archaeological finds and human camps), estimate the population of the pre-cultivation epoch at approximately the same size – 2–3 million people. These estimates refer to the present human – Cro-Magnon (*Homo sapiens*) – who appeared 100,000 years ago. Palaeontology has shown that the lifetime of the species averaged 7 My. With respect to this estimate, one can say that humankind has lived only a negligibly small part of his possible lifetime – less than 0.014%. Each of us has the same genome as the first Cro-Magnon.

Large animals have a limited memory – a perception and ability to process environmental information based on a programme of positive and negative emotions. The need for memory is determined by the fact that information within the environment greatly exceeds the "holding capacity" of the genome of any organism. Therefore, only a sensible strategy of behaviour can be recorded in the genome based on invariability of the average characteristics of an ecological niche. Then, at contact between the animal and environment, a behavioural tactic is formed – its choice being based on positive and negative emotions. This information is then fixed in extra-genetic memory in the form of conditional reflexes and imprinting (information imprinted on the memory at an early age over long time periods, such as the coordinates of the place of birth of migrating birds or fish, and human speech). Other animals do not pass this information on to their progeny, but humans do, and do so without the participation of the system of positive and negative emotions. This had determined the formation of modern man: differing from his precursor – Neanderthal man – through a well-developed speech apparatus. Presently humans are not changing genetically, since the rate of technological evolution, which is termed "progress", exceeds the rate of biological evolution (Gorshkov, 1995).

The Neolithic (agricultural) revolution had not only accelerated the technological revolution but also changed the character of population growth.

Before the time of cultivation the population had grown according to the availability of new territories, thus restricting a dramatic growth of population density. Violation of this veto was regulated by natural laws: upon consuming all the edible biomass in a given territory, humans needed to move to new territories or die of starvation. The appearance of agriculture cancelled this prohibition. It determined a settled life for crop growers and, in principle, for stockbreeders. The agricultural revolution also ensured an accelerated cultural specialization that triggered the formation of civilization.

Cultivation made it possible to increase the share of vegetable products in the human diet. From a single plot of land, humans could now feed on vegetable products, whose mass constituted a substantial share of the mass of pure primary production from a similar territory, the area needed to feed one human being 1,000–5,000 times greater. As soon as human beings began to consume arable land produce in such amounts that natural self-reproduction was impossible, he started to cultivate, fertilize, and sow arable land. The resulting additional food products stimulated human population growth on the Earth. Some 6,000 years after the Neolithic revolution the population reached 40–50 million, and by the beginning of our era estimates from different sources put the population at 100–200 million. This is therefore seen as the time when rapid population growth began. This point can approximately be represented as the year 1492 – the time of the discovery of America – when the population constituted about 0.6 billion people. A population of 1 billion was reached at the beginning of the Industrial Revolution, between 1800 and 1850. Between the beginning of the Industrial Revolution and the year 2000 the population has increased to 6 billion.

The population grew despite local ecological, resource, and social crises as well as epidemics. Since the main growth of the population prior to great geographical discoveries had been in areas of propagation of agricultural civilizations, all crises had taken place in these areas. Polibius (about 200–100 BC) wrote about feminine sterility in Hellas, reduction of populations, depopulated towns, and crop failures, though at that time there had been neither wars nor epidemics (Wallon, 1941).

Downfall of Rome and the ecological–resource crisis

The downfall of the Roman Empire at the beginning of our era was also connected with the reduced fertility of soils, low level of agriculture, and the efficiency of slave labour. The countryside became depopulated, in the 4th century in Campania and Sicily 500,000 ugers (1 uger = 2,942 m^2) of land was desolate. The resource food crisis and malaria undermined the power of the Roman Empire. It is possible that in searching for new territories the hordes of barbarians, Huns, Goths, and Vandals invading Europe from the east had also escaped from the ecological–resource crisis resulting from pasture impoverishment and, probably, natural disasters (Wallon, 1941).

The great geographical discoveries had saved Europe from resource crisis and overpopulation, and postponed social upheavals. But the upheavals did take place before the beginning of the demographic transition, when masses of young people occupied all spheres of life. The French Revolution was caused by a resource crisis intensified by two years of crop failure. Most of the leaders of the revolution were either young people or of middle age. The Industrial Revolution, with developed technologies of sanitation and hygiene, had determined very rapid growth of the population over the world. A single species – *Homo sapiens* – possessing additional energy from fossil fuels (and now nuclear energy) began to propagate over the entire planet to provide for his lifestyle, ousting other species, which could not compete with him. Humankind made them extinct and to this day is still destroying the system of life that has existed for 4 billion years.

The actions of the human hunter did not always coincide with his instinctive aspirations driven by genetic programmes. In brand new conditions, in an artificial habitat, the frequency of occurrence of such inconsistency increased drastically, and instinctive aspirations strongly modified by human intellect turned out to be more and more directed to an achievement of other goals, far from those intended by nature. Under the new conditions of the artificial environment some genetic programmes, not supported by actions, started disintegrating. Other programmes always supported by respective actions were preserved, but gradually they were transformed into intellectual stereotypes of thinking, cut-off from genetic programmes. For instance, this happened with the human genetic programme of violating the environment inherent in all large herbivorous animals akin to humans. It emerged in the form of deliberate unlimited destruction of ecosystems of mainly wild animals, in the name of progress. A hyper-aggressive attitude to nature developed, expressed in the stereotypical "technological optimist" with the slogans "to conquer nature", "to develop a territory", "not to wait for favours from nature but to take them", etc.

The ability to extragenetically convey information to descendants, a creative element, a desire to analyse and estimate, and other intellectual abilities of man have prompted him to develop new technologies. They, in turn, became a stimulus for discoveries and inventions, generating improved technologies for extracting and consuming natural resources. This process was constantly advancing, and in the Hellas period reached a very high level. It was then that the belief in scientific knowledge of nature appeared, as well as conviction that knowledge is power, which one can use to dominate nature, and control and direct the ecological processes. The Hellas world became rational, based on a certain sum of scientific knowledge in different subjects. This was used to subdue not only the "lowest" elements of nature (lacking intellect) but also the barbarians not possessing the knowledge of Hellenes and Romans.

After the Dark Ages and the beginning of the Middle Ages, rationalism underwent a revival. Francis Bacon, Galileo Galilei, Isaac Newton, Adam Smith, and french encyclopaedists had made a huge contribution to this revival. Based on scientific discoveries and creation of new technologies, man transformed his world beyond recognition. He utilized fossil fuels, organized mass production, continuously developed and modified systems of communication, created a universal system of education and management, etc. The 20th century became a period of the highest growth and propagation of rationalism. Man finally believed that through intellect (scientific–technical progress, reasonable political and economic measures, etc.) he would be able to solve all problems. Modern man rejected God, returning Him to the heavens, believing science was the new religion. He decided to establish his supremacy and develop nature and society on the Earth at his discretion.

Global and regional religions are now split into sects, while new and synthetic religions have been created that equate to atheism. Man monopolized the right to assess everything and formulate ideas of morality, though the latter is, in fact, more inherent in animals whose relationships are regulated by a strict system of both inherent and acquired ethical orders, which for man have been considerably eroded (Krasilov, 1992). An example of which is the expression: "no sin – no confession, no confession – no salvation."

In separating himself from nature, man separated his intellect from his biological nature. Intellect, instinct, and emotions diverged. Man has been suffering from this divergence for millennia. It has grown with developing technologies and growing urbanization, and has been especially rapid during the last century of scientific–technical progress. Krasilov (1992) called this division (separation of rational (intellectual) and natural (genetic) elements), "a chronic schizophrenia".

Both written and unwritten laws of modern society do not take into account human genetic aspirations. So, the existing norms and socioeconomic organizations of society sharply reduce the ability to maintain in society the genetic programme of preserving the normal genome. Real relationships between countries, nations, groups, and individuals based on intellect constantly lead to a violation of the genetic veto on murdering their own kind. Both at a governmental and a group level, an image of the enemy is constantly generated and reproduced. This led to the development of the most perfect means waging wars, organizing acts of terrorism, and criminal murders. Usually such actions are justified by "progressive ideologies", "true religions", "nationalism", "economic interests", "rights and liberties", or simply "vital interests". All these "rational" grounds are used with equal success by countries, individuals, and terrorists.

At all levels – domestic, national, and political – false information is constantly generated and used by individuals, groups, and even nations for misguided actions. This further erodes ethical principals. In the 20th century, the bloodiest wars took place, the most fearful weapons were created (nuclear, chemical, biological, and laser), unusual mass inhumanity was demonstrated, terrorism became widespread, local wars were constantly waged, as were economic wars, and all of this while the barbaric destruction of nature was still ongoing.

So, the current ethical norms established by man himself differ from his actual behaviour, in the same way as the behaviour of a religious man differs from the dogmas of religion. At the same time, the actual behaviour of man often contradicts his genetic aspirations. This duality of standards and gap between intellect and natural elements represents both the moral and ecological crises of humankind generated by modernistic rationalism.

The present rationalism, within which humankind cannot solve social problems, is based on aspiration for expansion (territorial, economic, ideological, political, and cultural) and domination over nature. Man rapidly "eats up" the ecological (and other) resources, depriving future generations of these resources and, therefore, the right for a better future. *While creating powerful technologies, defeating many diseases, and solving the problems of cloning,* man remains indifferent to the living beings of nature and ruthlessly destroys the foundations of his own life – the biosphere. The final crisis of civilization results from man's spiritualism, *his incorrect assessment of himself, the chosen vector of civilization development, and his relationship with nature. It does not mean that man loves himself, but rather his weaknesses and faults.*

The ability of man to create extragenetic information and pass it on to his descendants has separated him from nature and formed in him new egocentric ideas of the world. Man has turned the biosphere into a simple larder from which he takes resources without regard for the consequences, believing resources to be inexhaustible.

5.2   DELIMITATION AND ASSOCIATION

The precursor of modern man had triggered the process of globalization, when he left his ecological niche in the tropics and subtropics, and used technologies developed on the basis of extragenetic information, allowing him to overcome the ecological and biological boundaries. This process ended about 30,000 years ago. As the archaeological data show, in the Paleolithic period – a period of the population of the whole planet by humans – the cultural monuments had been uniform. But already during the Mesolithic (16–12,000 years ago) and Neolithic (12–7,000 years ago) periods, different monuments of material culture appeared, characteristic of certain places on the global map of human population. By this time, Cro-Magnon man had taken the place of Neanderthal man, and it was he who populated America and Australia.

The appearance of differences in the monuments of material culture denoted the beginning of a new global process: delimitation in space and in time of human populations, separation of individual groups, and the formation of various languages and communication systems. The separation of cultures took place both in space and in time, which led to different rates of progress between populations, formed differences not only in the material culture but also in intellect, ethics, relationships both within the communities and between men and women, as well as between adjacent communities. And all this was affected by the surrounding natural conditions. Even diseases were different for separated populations.

Small isolated populations

Many populations, isolated in many cases right up to the end of the 20th century, retained their population size, often being very small. Individual isolated tribes in the jungles of the Amazon, on the Philippine Islands numbered only a few dozens of people. Usually, these were communities with the material culture of the Stone Age, which followed the K-strategy (i.e., they preserved a stable size and density of population over their territories). Thus in small isolated populations, progress was extremely slow. It was directed mainly to an improvement of tools and methods of obtaining food.

Australia, discovered in 1770 by the English explorer James Cook, was populated by tribes which did not know of metal or the wheel. The bow was used only in the very north-east of the country. Australian aborigines were hunter-gatherers and led nomadic lives. They numbered about 250,000 to 300,000 people and populated the entire continent. The population density was 3.5 persons per 100 km^2. Nevertheless, aborigines had active contact between the tribes, exchanging tools and information. Tribes tended to have cultural specialization of their own. Some of them made shields, others boomerangs and axes (Velichko and Soffer, 1997).

Natural conditions in Australia (mainly deserts) had determined a very low population density and isolation of tribes, which in turn split into individual groups with fodder territories of their own. This did not favour rapid progress. Australia turned out to be the single continent where aboriginal man had not managed to develop agriculture and stockbreeding. Only dingos had been partially domesticated. Apparently, this was also connected with the fact that Australia was not the centre of the origin of cultural species of plants studied and classified by Vavilov (1926).

Most of the territory of both Americas, before their discovery by the Europeans, had remained populated with tribes living in the Neolithic epoch. They were divided into a lot of tribes with languages and dialects of their own. Such groups in America numbered more than one thousand. Some of them specialized in hunting, others in fishing. There were also primitive gatherers. At the same time, two centres of tropical agriculture appeared in America. The most ancient of them coincided with the Peru–Ecuador–Bolivia centre of the origin of cultural plants selected by Vavilov (1926). The agricultural centres appeared here much later than in the Old World – 7,000–8,500 years ago – the first cultural plants being vegetables, not cereals. Another centre of agriculture appeared not earlier than 4,000 years ago in the centre of Mexico. Over huge outlying areas of these comparatively small centres of agriculture and stockbreeding roamed tribes of gatherers, hunters, and fishermen (Bondarev, 1999).

The territory of the Old World was developed in another way. These lands had been populated much earlier than America and Australia by groups of humans from Africa. In newly populated territories, due to improved methods of hunting and gathering, the size of the population grew and reached a high density. The cave drawings of that time depicted clashes between different tribes. However, there were not only clashes but also trading with other peoples, as with the Australian aboriginal. On denser populated lands there were contacts with different cultures and an exchange took place of objects, food, and knowledge. In the regions of old populations (Mediterranean, Near-Asia, Indian Peninsula, China, Ethiopia, and Central Asia), which coincided with the centres of the origin of cultural species of plants (after Vavilov 1926), both agriculture and stockbreeding began to develop and propagate on the basis of a long history of gathering.

The history of these regions is a chronicle of attempts to combine the world known to people at that time – Oikumene. In the Mediterranean region the most ancient states appeared more than 6,000 years ago. Apparently, they originated as small territories gravitating towards some centre – an urban settlement. Such were, for instance, the first states in the Euphrates–Tigris region and in the valley of the Nile River. But over time, such settlements began to merge and great states were formed. Historical data show that as far back as the First Dynasty the Egyptians organized military and trade expeditions to Mount Sinai and to Ethiopia.

Similarly, the states in the Euphrates–Tigris region were formed, where one of the several settlements dominated all others, as was in the case with Ur domination. In the same way an associated state in other territories was integrated by means of military force.

Greek towns/states remained isolated for a long time, although the threat of the common enemy (the Persians) made them become allies. These towns/states favoured an association of population and territories around the Mediterranean through trade and creation of colonies – similar towns/states. Expansion was not driven by a single motive. However, the main motive was local land impoverishment and ecological catastrophes caused by the broadening of agriculture because of the advent of iron tools between the 7th and 3rd centuries BC. Ploughing up slopes and forest felling had caused erosion processes and impoverishment of land resources. Plato (427–347 BC) wrote that rocks appeared in place of the grass-covered hills and at one time a stone plateau had been fertile arable lands. Greece became an importer of timber and wheat. Therefore, Greek towns diversified their economics, began to trade articles of craft industries, wine, and olive oil. A forced emigration began from the country. This created, in foreign territories, port towns.

Phoenicians followed in a similar way, organizing their settlements in the Mediterranean, in particular, Carthago on the northern shore of Africa (present-day Tunisia), which became an important producer of wheat. The reason for this activity from the Phoenicians was the same as for the Greeks: forests in Minor Asia had been eliminated earlier than in Greece. Lands gradually became deserts.

Thus, the processes of delimitation and association took place in the ancient world. Processes of association began at the appearance of agricultural technologies. Where they appeared, states were formed striving for expansion. As a result, large territorial associations were formed, usually encouraged by military force, including a considerable part of the world known at that time. The first global association of the ancient world was the Empire of Alexander of Macedonia (356–323 BC) – the most ancient monarchy.

Ancient Rome became another similar association after its victory over Carthago (146 BC) becoming the greatest Mediterranean power. The Roman Empire reached maximum dimensions in the second century in the days of the Emperor Trajan. In 395 it was divided into western and eastern Byzantium, and in the 5th century the west Roman Empire disintegrated completely.

The ecological–resource crisis in the Roman Empire

Local ecological violations, mainly deforestation, were one of the causes of the downfall of the Roman Empire. The shortage of timber on the Apennines became apparent in the period of the Etruscans. Gradually the land degraded. Small farms could not compete with large latifundia, which used slave labour and oriented themselves towards the market. However, slaves were also not interested in maintaining the fertility of lands. One of the Roman specialists in agriculture, Columella, wrote: "We give agriculture, like to an executioner for massacre, to the most worthless ones – slaves" (Bondarev, 1999). In the 1st century, yield capacity started to decrease with 0.46 t per ha being considered a good yield. The area under cereal cultivation decreased. Those areas were replaced by pastures, about which one contemporary wrote that pastures ousted all other cultures and farms were transformed into deserts.

Later on, large empires time and again sprang up in the Old World (e.g., the Arabian caliphate and the Mongol Empire). China associated and disintegrated several times into individual kingdoms and principalities. But, in the final analysis, all empires created by force always break up, as mentioned above, and often do so as a result of local ecological violations. Global processes of association also took place in other areas, mainly technological and ideological.

The Greek and Phoenician colonization, marches of Alexander the Great, and the conquests of the Roman Empire all led to the propagation of agricultural technologies over the known world, that is, to a gradual creation of a single technological–economic basis for the greatest portion of populated regions of Eurasia – useful for agriculture and stockbreeding. The Romans were especially successful: when there was a deficit of wheat in Rome, it was brought from Carthage and Gallia (France).

An especially intensive development of agriculture in Europe took place during the 11th–13th centuries (see Section 4.3). This period of "the great uprooting" led to a sharp reduction of forested areas in the 15–16th centuries. Spain, England, the Netherlands, and France began to import ship timber, and as a result, metallurgic plants, smithies, and salt works faced an energy crisis since the basic source of energy and technological material in metallurgy was charcoal.

In that period the climatic conditions in Europe worsened and the so-called Little Ice Age set in: winters became cooler and longer, rivers started freezing, and in England the vines brought there by the Romans could not ripen. The 14th and 15th centuries, over most of Europe, turned out to be the wettest for the last 2,000 years. It was in these centuries that military conflicts began: The 100 Years War (1337–1453) and the Italian wars (1494–1559) between France, Spain and the Holy Roman Empire. Social situations worsened drastically. The American historian Walter Webb (1964) wrote that 100 million people lived in Europe in the year 1550. This community was poverty-stricken and starved. The way out of this complicated socioeconomic and ecological situation was the discovery of America and then Australia. As Webb wrote further, the Europeans obtained a vast territory 5 times greater than that of western Europe. Its development caused a socioeconomic boom in Europe, which continued until 1890, and this saved Europe from the "Malthus Trap" (overpopulation trap) (Arnold, 1999).

Development of new lands significantly contributed to the economics and culture of Europe. Another contribution was less apparent – historians do not mention it. It was the process of globalization of agricultural technologies, which, after great geographical discoveries, became a planetary phenomenon, like the ecological consequences of this globalization. These agricultural technologies, cultural plants, and animals moved from continent to continent, the agriculture of America developed like that in Europe. An exchange of plants took place between the tropical agriculture of Africa, Asia, and America. An intensified exchange between continents, earlier isolated, had serious ecological consequences: as a result of introduction and invasion of cultural and wild species of plants and animals into new ecological conditions, the latter often acted invasively, especially if in these new conditions there were no natural predators, parasites or pests.

The situation within human ecology had also changed. The Plague – "Black Death" – was brought to Europe from Asia (1346–1351), killing between 25 and

50% of the population in Europe. With the discovery of America and Australia, newcomers from Europe brought diseases previously unknown to populations in these regions of the world. These diseases took more American Indian lives than all the wars waged against them. Some authors call it the "holocaust" of the New World, others ecological "imperialism", since the population of the New World decreased from 75–100 million to 250,000. But the Europeans brought to the New World not only their diseases, but also slaves from Africa carrying Yellow Fever and other diseases. The New World reciprocated, giving some disease to the Europeans and the rest of the world. This exchange resulted in the formation of a united fund of diseases of the whole global human community – one more consequence of globalization.

The industrial revolution started in Europe in the 19th century, and rapidly began to spread its technologies over the world, which had already been acquainted with agricultural technologies and products exchange. But while it took almost 10,000 years for agricultural technologies to become a planetary phenomenon, industrial technologies took only 200 years, and information and communication technologies only 20 years.

Nowadays, the materialistic culture of the world is widespread, and the common material space has formed common views and morals. The common material space is maintained and provided for by common education. Therefore, it is ironic that American universities are places where Russians and Jews from Russia teach Chinese, Indians, Negroes, Arabs and Americans natural sciences and mathematics.

Another sphere uniting people is ideology. Man is characterized by an aspiration for creating ideologies, doctrines, and religions. In the period of delimitation of human populations both before the Neolithic revolution and after it, isolated communities and newly formed states had gods and patrons of their own. If sometimes they coincided, it was explained by a small choice or insufficient fantasy of their creators.

Gradually, more general religions began to form, which became the possession of not only the people of one group or state but also of different groups and states. Such a religion was, for instance, the Hellenic pantheon, assumed by the Romans and some other nations. Then a desire appeared to reduce both the number of gods in each religion and the number of religions themselves in order to more efficiently unite people. Over a period of about 1,000 years, beginning from the 5th and 4th centuries BC to the 7th century AC, three of the greatest world religions appeared as a result of the competitive interaction with earlier faiths: Buddhism, Christianity, and Islam. One could add to them Confucianism, which is also an ethical doctrine.

Morals are closely connected with religion. Observance of morals raises a man's prestige. However, at the early stages of the formation of the world religions, processes of delimitation started, often determined by national, group, or political motives as well as by the struggle for power within religious hierarchies.

Religious dogmas of Christianity slowed down the development of science and engineering. Christianity stopped the development of ancient science and engineering. Galileo Galilei and Geordano Bruno both exemplify the conflict between science and Christianity. One of the outstanding scientists of the Middle Ages – Leonardo da Vinci – had to record his developments by cryptography.

In China, where many outstanding discoveries had been made before our era, Confucianism had also slowed down scientific–technical progress, since it had created a system of values in which primary importance was given to stability determined by a strict hierarchy, rituals, established order, and respect for traditions of the past.

This slowing down of the scientific–technical progress had provided humankind with a 1,000 years of existence without collisions with nature, within the economic capacity of the biosphere.

Now most of the countries stick to the so-called freedom of conscience, which means a simultaneous professing of various religions and hence different morals. In the authors' opinion, this leads to a religious chaos full of negation of the existence of useful religious information – equating to atheism. In this situation, various religious communities always appear, which unite a lot of people who want neither to think nor to seek scientific verifications and experimental facts. It is easier for them simply to believe. As a rule, some people transform such communities into instruments to achieve their own selfish or imperious aspirations, and then these communities are transformed into totalitarian systems dangerous both for people and the state. Practically none of the world's religions prohibit the violation of ecosystems and abuse of the environment. Moreover, there were many who, after God, put man at the centre of the world. *Thus, religion and it's associated morals have never prevented man from destroying nature.*

5.3   LAG AND SURPASSING

Mastery of the planet by man has led to a separation of populations into individual groups of people and formation of new cultures with marked differences – in place of the single material culture of the Paleolithic period. Different languages have appeared as well as ideas about the surrounding world. At the same time, all groups of people living in the Mesolithic and Neolithic epochs and modern nations have a common origin: their precursors had been Cro-Magnon man, who populated America and Australia. They ought to have seemingly developed more or less uniformly, but the history of civilization shows that conquering of the planet by man had broken the rate of progress not only in space but also in time. This gap had appeared in the Neolithic period and was marked by transition to a brand new technology of food acquisition – agriculture.

Apparently, the Neolithic revolution was not restricted to one area. At least four regions have been recognized as ancient areas of propagation of certain agricultural plants. In each of these regions, agriculture had begun independently of the other regions.

Considering cereal crops: in Egypt and the Near East these were wheat, barley, oats, and lentils; on the Indian Peninsula, sorghum, and rice; in China, millet, kaoliang, and buckwheat; and in America, maize. In America, cereals were cultivated much later than the other regions. This suggests that most ancient centres of agriculture appeared independently of each other. In the Old World these centres were located between the tropic of Cancer and 40°N, while in America they were between the tropic of Cancer and the tropic of Capricorn. Another agricultural centre to consider was that in Mexico, which appeared much later than in South America. Archaeological data show that the most ancient centres of agriculture were in Asia and the eastern Mediterranean. The latest findings, testifying to first indications of agricultural technologies, show that these areas are 11,000 to 12,000 years old. Information has recently come to light about ancient centres of agriculture in north-western Thailand and in Indo-China 6,000 to 10,000 years, but this scatter of dates is too wide. It is important to note that the first crop growers appeared in most ancient settlements, since the flow of settlers initially had been from Africa to Asia and further to the east (i.e., without leaving the relatively warm belt of the northern hemisphere).

In all the regions of the world, after the Neolithic revolution progress began to accelerate: cultural specialization rapidly developed, towns appeared, building science developed as well as sciences connected with agriculture, astronomy to determine the times of agricultural seasons, and geometry of the Earth's surface to measure land sites and yields. Finally, writing and counting developed, first towns were built, trade developed, a system of management appeared, and military organizations and states were formed. All this required a broadened food production and development of new technologies. The process began through extensive economic growth.

In the period of hunter–gatherers, people's life expectancy was 20–25 years, hardly time to leave any descendants. Approximately half of these died without reaching puberty. Agricultural technologies favoured an increase in life expectancy. Infant mortality reduced and, as a result, the life expectancy increased by 5 years compared with the Paleolithic period. Populations began to grow due to both developing new territories and at the expense of obtaining additional produce. The trade between bordering states and wars led to a closer interaction of cultures and exchange not only of products but also knowledge, which favoured an acceleration of progress.

As far back as the period in the agricultural practice when stone tools were used, human beings had invented the plough, axe, hoe, and hand mill. They had mastered spinning and weaving and built reed and clay huts. In the Bronze Age there appeared sailing boats, the wheel, and roads. Stone tools were replaced by metal ones. Riveting and soldering, as well as metal vessels and weapons became customary. Brick, stone, and wood became building materials. Furniture and glazed earthenware appeared. All these technical means led to the formation of an artificial habitat and the cultivation of agricultural fields. People built dams, irrigation canals, and towns, expanded stockbreeding in order to use cattle not only as food but also as draught animals. Population size increased. Ancient Egyptian drawings dated 2565–2258 BC depict ploughing assisted by antelope. A drawing of ploughing with the use of draught oxen in Mesopotamia is dated to the third millennium BC. And one of the early drawings of scenes of milking is dated 2600 BC. All these achievements spread rapidly over all regions of ancient agriculture. Only in South America was development more slow. For instance, the wheel was not utilized, although the production of bronze, copper, and gold had developed (Bernal, 1956).

About 3,500 years ago the Iron Age began, with which many important achievements were connected: the alphabet was invented, coining was started, philosophy was formed as well as foundations of rational science. The last millennium BC was the golden age of the ancient culture.

Ancient culture reached a very high level and had its roots in the achievements of the people of Egypt and Asia. In the countries of the Near East, Egypt, India, and China, despotism and patriarchal-stagnant traditions always dominated, which restrained the competitive interaction and aspirations for creativeness, genetically inherent in man. Everything obeyed a certain order, traditions, caste restrictions, and the state. It was most important to preserve the old order, its reproduction, and stability. All instructions were sent out from above, the rudiments of scientific knowledge were transformed into orders for practical use, and religions became an instrument to preserve the system and its ideology. All this effectively excluded any competition.

In ancient Greece, after overthrowing tyrants, society moved on to another form of government involving masses of people with rights to have opinions of their own and with rights to vote. Changes also concerned property, which in Greece became more and more independent of the state, in contrast to earlier despotic communities, where property (estates) belonged to the government, which was free to redistribute them as it saw fit. Changes in the political system from despotic to republican and in the system of property ownership resulted in a new effective system, which included competition in all spheres, including science, arts, management, production, trade, and sports. Many authors call these historical events a kind of "mutation", which in our opinion is incorrect. It was a natural process of man's changing economic activity connected with ecological changes on the Balkan Peninsula as a result of intensive agricultural land use. Felling of forests, ploughing of slopes, and breeding of goats and sheep all led to land degradation and agricultural decay, which forced the Hellenes to diversify their economy and change its structure by aiming it towards export in exchange for wheat. For instance, Athens imported cereals used to feed 300,000 people of the town. Regarding agricultural plants, only olive trees and vineyards were preserved, therefore the Hellenes from the parent state exported olive oil and wine as well as earthenware, weapons, and other articles of craft industry. This economy differed from pure agriculture and therefore required other forms of management.

During meetings, people disputed the justifications of various views of the problem. It was not a dispute of authorities whose quoting was not considered a serious argument. It was via Greek mathematics that knowledge was put in the form of theorems: "given – to be proved", whereas in the mathematics of ancient Egypt and Babylon solutions were put in the following scheme: "do in this way – you did it correct".

The Romans assumed the Hellenic culture and transformed it to fit their society. The Greek–Roman social system emerged from the world at that time with traditional patriarchal norms of life often called "the East". Both Greece and

Rome demonstrated scientific and technical achievements, rapidly progressing science, rationalism, freedom of personality, and protection of ownership.

Respectively, if after the Neolithic revolution, agricultural civilization developed much more rapidly compared to the rest of the world, about 2,500 years ago a new variety appeared within this civilization, which differed, first of all, through a greater state independence of the owner, personal freedom of citizens, and rapid progress of science and engineering. The resulting scene resembled the modern world: Hellas and Rome – modern developed countries, "eastern despotism" – the second-world countries, an analogue to the former socialistic camp or countries with transitional economies, and, finally, the rest of the world – the third-world countries, most of which still lived in the Stone Age.

The Roman Empire, which replaced the Roman Republic, transformed gradually into an ordinary "eastern" despotism, though the technical and scientific progress within its boundaries still continued. At the start of the Dark Ages the great migration of people from the Asiatic steppes led to many achievements being lost. Barbarians established typically eastern norms of life. From this time on, everything connected with community works had been forgotten. Communications, irrigation systems, and water supply were lost, and trade with many countries ceased. Earlier forms of cultivation returned: a supreme sovereign granted land to his vassals for their service, and those in turn used the peasants' labour on the land, also giving them land sites and obtaining from them rent, taxes, and corvée.

The intrusion of barbarians into Europe, among other causes, was connected with the desiccation and desertification of the belt within which the first centres of agricultural civilization had developed. Numerous facts show that during the last 7,000 years this process remained unbroken despite climatic fluctuations (i.e., it was determined by destruction of forests and other natural ecosystems over huge areas and their replacement by agrosystems, leading subsequently to desertification).

The beginning of a new era was marked by the propagation of organized religious faiths: Christianity, Islam, Buddhism, and Hinduism. It was at this time that their doctrines were formulated and clergy was organized. All these religions are characterized by hierarchy, regulation of rituals, and a dogma as a criterion and uniting element that included blind belief in the arrangement of the Universe described in sacred books. All this corresponded to the mentality of the eastern despotism, and therefore suited the form of power existing at that time.

In the regions propagated by Christianity, first of all, in Europe, intellectual life and science were subject to the Christian dogma and more and more immersed in the clergy circles. Impeding of progress in Europe reached a maximum in the so-called "Dark Ages" (4–9th centuries). In other continents, where such rapid progressive changes were not observed – India and China – peaks of cultural development occurred. Periods of economic and cultural peaks within the eastern despotism with eastern types of land property in India were connected with the empires of Gupts (320–480 AD), Chalukyev (550–750 AD), and in China with the dynasties Wei (386–549 AD) and Tang (618–900 AD). At that time there was also an increase in Islam.

In the 8th century the Arabs invaded lands from India to Spain – former possessions of the Roman Empire in Asia (except Asia Minor), North Africa, and Spain. Over most of this territory, a single culture, religion, and literature was established. Similarly, during several hundreds of years there was a single administration. Science developing here used the achievements of both Hellenic and more ancient cultures – Egyptian and Babylonian. In particular, in India an improved system of counting was invented with a certain order of numerals and with a zero – which we now call Arabic numerals.

In the 13th century, under the pressure of the barbarians (Mongols and Turks) and as a result of the inner discords and decline of agricultural production, the Arab caliphate and the Byzantine Empire fell – barbarians defeating both China and the Indian states. In these regions the development of science and society had been broken for centuries.

After the 9th century the golden days of Arab science were over. While in some sciences Arab scientists surpassed the ancient ones, in philosophy and understanding of the world they were inferior to them. Despotic forms of government and dependence of proprietors on the tyranny of the state had not emancipated the people's thoughts. Strict hierarchy and dogmas of religion kept people within rigid limits and forced them to promote the same system and strictly follow traditions. The scientific achievements of the Arab world were not lost, they were assimilated into Europe along with Hellenic culture.

At this time, lands on the European continent were rapidly developed, and this continued until the 13th century. The two-field system was replaced by the three-field system, a wheel plough came into use, and winter crops appeared, which doubled annual yields. An expansion of crop areas and increased productivity ensured economic and demographic growth. In England, towns built by the Romans revived. In Italy the half-ruined towns (1,000 years old) began to acquire a new face. Roman basilicas and Gothic cathedrals were built. In France, during the period 1050–1350, 80 cathedrals, 500 large churches, and tens of thousands of parish churches were built. The warm and stable climate in the 12–13th centuries (in England, for instance, grapes were grown) ensured rapid expansion of crop areas producing a stable volume of food. During these centuries there were only 2 years of famine. By that time the population of Europe had increased 2–3 times. After the 13th century, the number of poor years increased sharply – averaging 15 per 100 years up to the 18th century (Bondarev, 1999).

The cathedral schools preparing the clergy became larger and more powerful to such an extent that some of them were transformed into universities with an established course of teaching the free arts, philosophy, and theology. From the mid-12th century, over 200 years, universities were founded in Paris, Bologna, Oxford, Cambridge, Padua, Naples, Salamanca, Prague, and Krakow. As far back as the 11th century, scholastic disputes occurred in order to reconcile sacred writings and works of clergymen with Hellenic logistics. Secular education and sciences gradually separated from the church. In Paris the Royal College was founded in 1530, where only humanitarian sciences were taught.

The social structure of society became more complicated, there was more legitimacy and freedom in managing private property. The growth of towns, development of market relations, trade, capital turnover, and association of small feudal estates into centralized monarchies led to a legalization of property relations between the king, feudal lords, and peasants. As a result, the king turned out to be the proprietor of only the king's domain, and free peasants had a right to sell their land (e.g., in England it was common practice already in the 18th century). Churches and church lands also rid themselves of the guardianship of the state, like many self-governing towns. The Renaissance and the Reformation triggered the process of formation of present-day democracies. The humanistic movement originated in Italy in the 14th century, and from there propagated to France and England as well as to other states of Europe, resigning hierarchy, establishing more secular relationships in society, and greater freedom of thinking. The Reformation, resulting in the appearance of Protestantism, rejected the ecclesiastical hierarchy, opposing clergy to lay men, simplified the cult, and rejected monasticism (i.e., made religion more democratic).

All indicated changes took place against important natural and economic events. From the end of the 13th century, the so-called Little Ice Age started, gradually worsening conditions of economic activity. Winters became cooler and longer, rivers (Thames, Rhine, Rhone, etc.) started freezing, and in England the grape vine could not be grown. Summer became rainy, and the North Sea was often stormy.

Plague in Europe

In 1346, fugitives from the Crimean Caffa besieged by Tatars brough plague to Europe, which propagated from Egypt to Iceland and the Baltic Sea, and farther to the west, to Spain. Disease rapidly propagated over vast territories over a period of 6 years (1346–1351) which testified to the fact that the size and density of the population in Europe had reached a critical level, creating "favourable conditions" for a plague pandemic. The 75-million population of Europe was reduced by 30%. In 1361 a new pandemic took the lives of many people born after the first one. In 1369, plague returned, taking 10–20% of the population. Every 10–12 years it returned until the 16th century.

Forest felling and the energy crisis in Europe

At the end of the 13th century, the possibilities for extensive development of agriculture were exhausted, but as a result of mass forest felling the processes of soil impoverishment developed rapidly. Forest felling also led to an energy crisis, especially in the regions of ore extraction and salt works. At the same time, in 1525 in Germany there were 100,000 mine workers. The total length of adits in Mount Falkenstein in the Tirol was greater than the length of the lines of the Moscow underground: 222 and 217 km (1998), respectively (Bondarev, 1999).

These natural, economic, and ecological events on the continent created conditions for a social crisis accompanied by food shortage, relative overpopulation, and the appearance of many younger people (average lifespan was 35 years at that time), who served as hired troops or became robbers, sailors, and pirates. The discovery of the New World (where flows of emigrants from Europe rushed, bringing with them the crafts and skills of European civilization) turned out to be a way out of this critical situation. However, from the New World the flow of silver and gold appeared – currency was devalued and prices jumped, marking the start of massive inflation.

By this time, due to Galileo Galilei and Nicolaus Copernicus, ideas about the Universe had changed: the Sun was put at its centre. Great achievements appeared in metallurgy – new metals were discovered and mass production of cast iron began. Chemistry and biology revealed man's structure and his blood circulatory system. Shipbuilding was improved. Progress accelerated, favoured by revolutions in the Netherlands and in England, where the merchants, craftsmen, and small farmers restricted the power of the king and landlords, giving way to a free market. The market turned out to be the best means of introduction of innovations, which stimulated the development of technologies and science. One of the creators of new ideas of the world was Francis Bacon who wrote that science should be aimed at enrichment of man's life with new discoveries and new power.

Again, Europe in its economic, social, technical, and scientific development stepped forward and from that time on it became ascendant. As in earliest times, the world community remained divided not only in linguistic, religious, and cultural spheres, but also in growing economic and social spheres.

Achievements of the Europeans taken to North America and later assumed by Japan, are based on ideas of inexhaustible resources of the Earth, of man's right to use them at his disposition, as well as of the power of science, with the help of which he can completely transform his surroundings and his planet. This is an ideology of modernism, an antropocentric world view being its main element.

Modernization triggered the formation of a new society. Industrialization and development of agriculture based on scientific–technical progress became the material basis of modernization. The main instrument was a free market. In the political sphere there was a transition to republican forms of government, democracy, and civil society. Man completely differentiated himself from nature, surrounding himself with the "armour of civilization".

In the social sphere, modernization changed relationships between people. Society became "atomistic" (Vishnevsky, 1998) as a result of the destruction of traditional agriculture with its close family and other ties between individuals. In the epoch of modernism, its dissident form appeared, which solved the problems of the modernization of society using the centrally directed system. But the conse-

quences were the same: industrialization, automization of society, natural destruction, neglect for the laws of the biosphere, and a conviction in the ability to solve any
problem by technological means, including the control of ecosystems and the
biosphere.

The best minds of the USA convened by the American Press Association in the
late-19th century demonstrated their optimism "predicting that most of the
problems would be solved and that the development of science and engineering as
well as the growth of material production would lead to a society which would
become almost a Utopia." The same optimism with respect to the on coming 21st
century was demonstrated in the journal *Business Week* (1998) in its special double
issue dedicated to the economy in the early third millennium. It predicted still greater
economic progress and the ability of the world economy to be able to solve all
conceivable social problems on the crest of technological progress. This view of
technological optimists stimulated by rapid growth of technologies especially
prevails in the mass media. This view contains a concept of human development,
according to which the development of human communities is independent of
nature. There is an opinion that "our economy based on informatization is able
to develop regardless of the global ecosystem of the Earth" (*State of the World*
1999, 2000).

The views of the authors of the journal *Business Week* could well be labelled
"self-satisfaction". In our opinion, they are mistaken in their belief that the independence of humankind and civilization from the global ecosystem – the biosphere –
is a new idea. This idea appeared 10,000 years ago and became the cornerstone of
modernism. Meanwhile, in 1944, Vernadsky stated: "... a human is usually spoken
of as an individual freely moving over our planet, who freely builds his history. So
far, historians and other humanitarian scientists as well as biologists deliberately
neglect the laws of the biosphere – the Earth's envelope, the only known place where
life is possible. Man is inseparable from it." After the man–nature clash in the 20th
century and development on the planet of a severe ecological crisis, his words
became most urgent. *The technological optimists in their assumption concerning the
future make two errors: they take into account only the achievements of the 20th
century, ignoring its huge outlay, and they forget a simple empirical truth – that all
predictions contain uncertainties, that man cannot foresee his future reliably enough,
and that future generations will have to pay for every mistake* (Kondratyev and Losev,
2002; Kondratyev et al., 2002a; 2002b; 2003a, 2003b).

6

The 20th century: time of confrontation with nature

6.1 GROWTH IN ALL DIRECTIONS

The planetary propagation of first agricultural and, from the 19th century, industrial technologies after the great geographical discoveries accelerated the trends to closer contacts between various countries followed also by a military–political association. For the population of Europe, most of the rest of the world turned out to be an "empty" space, which the emigrants from Europe rapidly populated. Aborigines were either killed or had to receive new settlers and their conditions. Several huge empires appeared: Spain captured most of Central and South America, as well as lands in Africa, Portugal controlled the territory of present-day Brazil and lands in Africa and India, and Great Britain held North America, Australia, then India and lands in Africa. France, Germany, the Netherlands, and Belgium did not lag behind. Russia developed Siberia and added the Caucasian states as well as those in Central Asia. By the beginning of the 20th century the world had been divided among European powers, and a small number of the remaining states fell under the influence of one of the European countries. Mastery of resources of the captured territories and their plundering favoured the industrial prosperity of Europe.

New masses of resources and populations were constantly involved in the process of production. An accelerated growth of the economy began. Based on industrialization, agriculture also changed due to a broad mechanization of works. Health care and medical technologies were rapidly developed and improved, thus reducing mortality, especially infant mortality, and prolonging the average lifespans not only in metropolitan countries but also in the colonies. As a result, populations sharply increased ("demographic boom").

But the economic progress in the 20th century was especially great. The global gross domestic product (GDP) in 1900 constituted US$2.8 trillion (in US$ of 1997), and at the end of the 20th century it exceeded US$43 trillion (*Vital Signs*, 2001). Table 6.1 gives macroeconomic estimates of the contribution to the economy of

Table 6.1. Gross domestic product (GDP) production and population size to the developed and developing countries, as well as in the countries with transitional economies in 1993. After Schlichter, 1996.

Activity	Global	Developed countries	Developing countries	Transitional countries
GDP production, trillions of US$ (%)	29.1(100)	19.5(67)	7.6(26)	2.0(7)
Population, billions of people (%)	5.55(100)	0.76(14)	4.16(77)	0.49(9)

various groups of countries in the early 1990s, which shows that 14% of the global population produce 67% of the global GDP. This means that in developing countries the annual per capita income averages US$1,827, in countries with a transitional economy it is US$4,827, and in developed countries it is US$25,658. In 2000 these estimates increased with the gap between countries increasing, too.

Energy is the basis of economies. As far back as the mid-19th century, the prevailing sources of energy were wood, charcoal, and straw (i.e., biomass). Only in the 1900s did fossil fuels (energy carriers) become the equal of biomass by volume of consumption. This led to the growth of the per capita energy consumption by a factor of 8. By the end of the 20th century the specific energy consumption increased by a factor of 20 compared to 1850 values (with account of increasing efficiency). In 1900 only 1% of fossil fuel was transformed into electrical energy, and by the end of the 20th century this had increased to 25%. In the period 1950–1995 energy consumption was at a maximum: its specific consumption increased from 400 to $2,300 \, \text{kWhr} \, \text{yr}^{-1}$ per capita. It is this enormous increase in energy flux that changed the world significantly in the 20th century (Smil, 1998).

Energy played an important role in the development of agriculture, which was, and remains today, the basis of the economy and existence of the world community, since it provides this community with food. The unprecedented development of energy use created conditions for a new "green revolution". From 1900 to 2000, taken from different estimates, the area of cultivated land increased only by one-third, whereas the yield capacity increased 4 times, and the total yield by a factor of 6. This happened due to an 8-fold increase of energy investment into agricultural production due to new agricultural machines and energy expenditure on their functioning, production of mineral synthetic fertilizers and means of plant protection, selection of new varieties of plants and animal breeds, as well as the use of vitamins, antibiotics, and mineral additions. From the time of the Neolithic revolution it represents the greatest increase in food production due to an increased yield per unit of agricultural area. Nowadays, to obtain one food calorie, more energy is required.

Present-day civilization is heavily dependent on fossil fuel burning. This process began in the Middle Ages in Europe when people started to use coal to heat and cook in the regions where coal strata outcropped. Since then the continent has been deforested. In London, coal was used for heating from the 14th century onwards. It had been extracted in Northumberland in the days of the Romans, but its large-scale use began only in the second half of the 19th century. During the 20th century global coal output increased from $170 \times 10^6\,\mathrm{t\,yr^{-1}}$ to $1.5 \times 10^9\,\mathrm{t\,yr^{-1}}$, with a peak in the second-half of the 20th century. The 20th century became the period of an intensive oil and gas consumption.

Oil and gas: history of use

Oil has been known for thousands of years due to natural welling in the Middle East, but it came into use much later. In Constantinople it was used to heat the therms (bath houses) in the days of the Roman Empire. Oil extraction started in the USA in 1850 in Pennsylvania, and in 1900 was extracted in Romania, Russia, and Indonesia. Nowadays about 30,000 continental and coastal oilfields have been discovered. Most oil supplies (four-fifths) are concentrated in five basins: in the Persian Gulf, Gulf of Mexico, Lake Maracaibo (Venezuela), the Volga–Urals region, and in western Siberia, with two-thirds of oil reserves lying in the region of the Persian Gulf. During the 20th century the oil output has increased from $8 \times 10^6\,\mathrm{t\,yr^{-1}}$ to $1.2 \times 10^9\,\mathrm{t\,yr^{-1}}$. In the last quarter of the century output has stabilized. Natural gas was known in China at the beginning of our era, when in the Sychuan Province it was used to evaporate salt brines. At the same place, 1,000 years later, gas wells were drilled, and via bamboo pipes, gas was brought to the surface. From the beginning of the 19th century, natural gas was extracted as a by-product of oil extraction. In 1900 gas output constituted about $500 \times 10^6\,\mathrm{m^3\,yr^{-1}}$, and by the end of the 20th century it exceeded $10^{12}\,\mathrm{m^3\,yr^{-1}}$ (Smil, 1998).

From 1956 a new source of energy started to be used – nuclear energy, on which great hopes were set. In 1971 a representative of the US Commission of Nuclear Energy, G. Seaborg addressing the International Symposium on peaceful uses of nuclear energy spoke about its wide use by the year 2000: he predicted that it will constitute half of the energy needed for the USA, it will drive both tankers in the sea and vehicles in space, and nuclear power will be used in the building industry. After his speech oil prices rose globally because of the supposed reduction of hydrocarbon resources, and it seemed that Seaborg's prognoses were quite real. However, the incident at Three Mile Island, the fall in oil prices, reduced energy requirements due to energy-saving technologies, problems of radioactive waste storage, and the Chernobyl accident hindered the growth of nuclear energy. Also, it became clear that shutting down nuclear power stations was a very expensive business. As a result, at the end of the 20th century, nuclear energy production peaked at a global production level of about $11,000\,\mathrm{TWhr\,yr^{-1}}$.

Growth of per capita energy consumption

To collect food, human beings used only the force of their muscles. In this instance human power consumption constituted 140 W. Hunters used fire and ploughing tools, increasing power consumption to 240 W. In the traditional agricultural society, the per capita power consumption reached 500 W. In our present society, the per capita power consumption averages 3,200 W, with developed countries, consuming more than 7,000 W per capita (Reimers, 1990; Gorshkov, 1995; Arsky et al., 1997).

Thus the 20th century, despite two major and bloody wars and the Great Depression, turned out to be most remarkable for the economic development of humankind. This was a century of triumph for modernism, its ideology being a market system and civil society, placing man as the centre of the world.

The 20th century was also unprecedented with respect to population growth (Table 6.2). In 1900 the global population constituted 1.6 billion people, by the end of the 20th century it reached 6 billion people.

In the same century, an accelerated growth of population was observed. At the very end of the century the increment ceased because of reduced birth rates in the developing countries (Kapitsa, 1995). A global demographic transition took place in the developed countries in the early 20th century (Figure 6.1). Total population in the history of humankind's development constituted 80–150 billion, with the average lifespan of about 20 years. Nowadays, lifespans average about 64 years: in developed countries 74 years and in developing countries 62 years.

Apart from broadening the possibility of providing people with food, the size of the population grew due to reduced mortality, especially of infants, and increased lifespans. These processes were determined by the development of "technologies to provide vital functions", which include a wide spectrum from building houses and clothes production to hygiene, sanitary, medicine, and recreational technologies.

The peak in population increase took place in developing countries where, throughout the 20th century, the "high birth rate–gradual reduction of mortality"

Table 6.2. Growth in the global population from the year 1800 to 2000.

Year	Global poulation (billions of people)	Time taken by population to increase by 1 billion (yr)
1800	1	–
1930	2	130
1960	3	30
1975	4	15
1987	5	12
1999	6	12

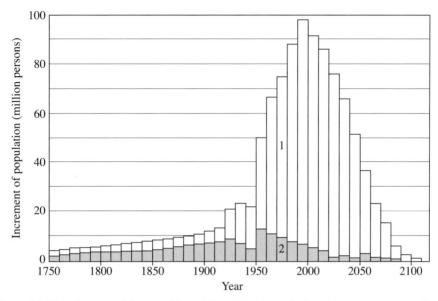

Figure 6.1 The demographic transition. The global population increment averaged over decades from 1750 to 2100. White – developing countries, tinted – developed countries.

strategy prevailed. In developed countries this strategy remained well into the 20th century, but then it was replaced by the strategy of "low birth rate–low mortality". Although in the first case the size of the population grows rapidly, in the latter case either the increase is small or the size of the population is stable. Sometimes, it even decreases (as is the case in Sweden).

Now, in most developed countries the size of the native population is decreasing, since in many countries (Germany, France, Spain, etc.) families have 1.5–1.7 children, on average, which does not ensure direct replacement of the population (for this to happen a family should have 2.2 children on average). However, the population of these countries has grown as a result of emigration from other countries. The direction of flows of emigrants has changed sharply: before and at the beginning of the 20th century, the flow of emigrants was out of Europe, while during the second half of the century emigrants arrived in Europe, the USA, and Australia from Africa and South-East Asia. This is one more distinguishing feature of the last century. At that time the developing countries reached a maximum population increment, and in the 21st century one should expect its rapid reduction. The latter has been manifested in several countries, where the birth rate has started decreasing.

In the 20th century the problem of family planning became not only a governmental problem but a global one. Some developing countries started to legally regulate the sizes of families, which were placing heavy pressures on economy and social spheres. These measures have been applied in China, Indonesia, and Thailand.

They markedly affected the population increment in these countries and the number of children per family rapidly approached 2 or slightly more.

The demographic boom has stimulated the growth of the economy and resource consumption. Nowadays about 300×10^9 t of substances ($50\,\text{t}\,\text{yr}^{-1}$ per capita) are extracted and transported in the process of obtaining raw material and final products. Final produce constitutes only a small part of this mass. Only recently, some countries began to perform a balance analysis of material flows. So, in Germany, from estimates for 1991, the inner material flow constituted $4.3 \times 10^9\,\text{t}\,\text{yr}^{-1}$, and the outer import $3 \times 10^9\,\text{t}\,\text{yr}^{-1}$, or $90\,\text{t}\,\text{yr}^{-1}$ per capita. Germany exports about 1 billion tons of products. In Austria, the inner material flow constitutes 117×10^6 t, imported flow is 40×10^6 t, and exported flow only 16×10^6 t. In Austria the annual consumption of water and air is 3.8×10^9 t and 330×10^6 t, respectively. Thus, with water and air taken into account, Austria consumes $4{,}287 \times 10^9$ t of substances.

On the whole, to obtain per capita final products, 50 t of substances, 800 t of water, and 3.2 kW of energy are transported and used annually. At the beginning of the 20th century all these indicators were lower than present-day levels by an order of magnitude. To obtain five metals (iron, copper, aluminum, lead, and gold), 45.7×10^9 t of ore are extracted every year, or more than 7 t per capita (Danilov-Danilyan and Losev, 2000; *State of the World* 1999, 2000).

6.2   THE DISINTEGRATION OF EMPIRES

The world used to be divided up into empires, the number of which was small at the beginning of the 20th century (see Section 5.2). The typical way of dividing the world up according to military–political action was done within the framework of what is now called "the West" or "the North". This took place 1,500 years after the fall of the Roman Empire and many hundreds of years after the fall of the Arab caliphate and the Mongol Empire. New empires differed from the ancient ones in that together they covered the entire world, had a limited set of religions, and one ideology – modernization. During the creation of these empires disruption, death, and cruel oppression had occurred. Gradually, the colonialism acquired more civilized forms, creating a system of education and introducing technological innovations, especially regarding health care.

Modernization in colonial India

From 1830, British engineers began to build irrigation systems in India. By 1890, almost 44,000 miles of main and distributive canals were built, which irrigated 13 million acres of soil. By the end of the British dominion in India, the length of irrigation systems increased up to 75,000 miles, and the area of irrigated soils – up to 33 million acres. The building of railways in India started in 1853, and by 1910 India had a railway network reaching 32,000 miles long. At the same time, the forests were felled. Nowadays, wild forests in India cover about 1% of the territory (Arnold, 1999).

For the local populations of the New World the initial periods of colonization were the most tragic, when they were cruelly robbed and dominated by force. Many died of the diseases introduced to locals. Two-thirds of the population of America died from such diseases (see Section 5.2). The huge flow of settlers from the Old World and ousting of the aboriginal people in the 19th century, resulted in a majority of the populations on the two continents of the Old World – America and Australia – either being from Europe or being descendants of Europeans.

The first colonies to seek independence were the territories on the eastern coastline of North America belonging to Great Britain. In 1776 they declared their independence and formed the United States of America. The next states to push for independence covered South and Central America, territories belonging to Spain and Portugal. By 1828, independent states were formed over this entire territory. In 1900, only Canada and Greenland belonging, respectively, to Great Britain and Denmark, as well as small territories in South America and in the Caribbean Basin belonging to France and the Netherlands, and Great Britain remained in America as colonial enclaves. This was the start of the disintegration of the empires created by the Europeans after their great geographic discoveries.

The final stage in the disintegration of empires fell within the last century, the first-half of which was especially full of military conflict. Humankind entered the 20th century fully aware of the need to have powerful armies. Owing to this "philosophy", huge empires were created. Upheavals, wars, and revolutions were standard in Europe in the 19th century, beginning with the Napoleonic Wars. Wars were started for resources. The strategy of dividing the world up between a small number of states did not suit other countries which considered themselves deprived of their due share of the spheres of influence. In the first-half of the 20th century this resulted in two World Wars – the bloodiest in the history of humankind.

The massive loss of life during these wars was determined by the development of military strategies that targeted civilians for the first time. The huge majority of victims, many of them civilians, resulted from the adoption of such military strategies. Most of the ideologies for creating a "beautiful" future, which appeared in the late-19th/early 20th centuries, lacked humanity. Supporters of these doctrines often encouraged terror and revolution. Of course, the age structure of the European population, with a high share of young people, played an important role. For instance, Napoleon at the age of 30 performed a coup-d'état. Inability to assess the consequences, lack of experience, adventurism, fanaticism, and extremism – all features inherent in youth – were factors that reduced humanistic elements. Moreover, the market system, which still depended on the "wild", "biological" market and still had not developed democratic institutions, made a substantial contribution to the formation of anti-humanistic conscience.

The First World War ended with the defeat of those countries that considered themselves cheated during the division of the world in the 18th and 19th centuries. Despite this defeat, a military philosophy remained in Germany. It did so in Italy and in the Soviet State, which formed after the First World War. For a long time the USSR was pressed and threatened by the countries with market economies – they were considered the enemy.

The Second World War was waged under the slogan of the struggle against the fascist totalitarian system. The opponents to fascism were the "western democracies" and the Soviet Union. However, these Western democracies were not true democracies, since many represented enormous colonial empires.

By this time, a national elite appeared in the colonies, with the aim of removing the power of the metropolitan countries. The social conscience of the people of the colonies generated ideas of independence, reinforced by the results of the Second World War. An important factor in this struggle was the appearance of large cohorts of young people who wanted to improve their position and ensure their future via independence. Therefore, along with the idea of sovereignty, they often raised socio-economic demands. Western countries in many cases tried to stop (by force) the disintegration of these colonies, but the world had irreversibly changed. Besides, many countries fighting for independence got help from the socialist bloc (especially if the struggle against colonialism was carried out under socialist slogans) or from the former colonies which had already gained freedom.

Gradually the western countries, as a result of a rapid economic growth and failures in war on foreign territories, began to understand that wars were unprofitable and that it is impossible to hold vast territories by force. Today, pragmatic and humanistic ideas prevail. Nowadays it is believed that it is much more profitable to wage economic and financial war in order to carry out economic expansions than to apply real military force. Within 30 years of the end of the Second World War all empires formed by European countries had fallen. There was only one empire left – the Soviet Union and the Soviet Bloc. This empire was also doomed, since the centrally directed system hopelessly lagged behind the developed countries in the race for technologies. The market turned out to be the best mechanism to advance innovations and to claim new technological solutions formed and selected in conditions of fierce competition. The only branch of the economy in the USSR that took part in international competition was the military. Therefore, only this branch developed successfully (suffering huge financial expenditures).

The Soviet Union lost both the economic and social races. Besides, the centralized system was already unable to control all sectors of production, and the system of economic management was incessantly confused. The country, in fact, completely depended on developed countries who bought Soviet oil worth US$70 billion annually, permitting the USSR to maintain its military–industrial complex, to buy grain in large volumes and high-quality consumer goods from the West. In 1991 the soviet system fell, leaving no empires left in the world. The end of the 20th century became a period of disintegration of peoples and states, the numbers of which increased three times compared with the number at the beginning of the century.

The colonial period in the history of many of the newly formed states has left a remarkable legacy. They have turned into Third World countries. Not only the people but also the natural environment has been transformed into a "colony". The territory of these states are nothing more than sources of raw materials and food for developed nations. This gives further meaning to the term "banana republic". New technologies were used to transform some former colonies into banana, coffee, sugar, and other republics, and some into oil, iron ore, and bauxite republics. Only at the end of the 20th century did some countries enter the period of modernization and creation of an industrial society gradually transforming into a civil one. The remainder are mainly used as raw-material appendages to the developed countries. Most of the countries of the third world have become, in fact, territories to which are allocated "dirty" resource-consuming products and waste disposal. *The developed countries, either having exhausted their own resources or attempting to preserve them for future generations, use the resources of developing countries and their ecological space. In this sense, colonialism has acquired a new face – instead of military–political it has become economic–ecological with the developed countries having the final say in the social make-up.*

6.3   URBANIZATION

The agricultural technologies resulting in excess food production, labour division, and the respective need of exchange and regulation, as well as protection of the resulting system, have led to the necessity to create centres of exchange and power. First, towns became such centres. Craft industries were also concentrated in these towns. Historical facts testify that an appearance of towns was connected with the agricultural revolution. In the territory of Australia, where before the Europeans, aboriginal people had been hunter–gatherers, there had been no towns. In North America, over most of its territory, where American Indian hunters lived, there had been no towns either. This was the case in most of South America. Towns existed only in small enclaves, where locals had reached the agricultural phase of development – in Mexico and in some regions of the Pacific coastline of South America. Thus at an initial stage of their development, towns had not been a global phenomenon.

The beginning of urbanization

First urban settlements appeared in three regions of agricultural civilization: in the Middle East, including the valley of the Lower Nile, in the valley of Indus River on the Indian Peninsula, and in China, in the basin of the Yellow River. The urban network expanded together with the propagation of agricultural technologies onto new lands. On the whole, it was a regional process: towns appeared in different regions of the Old World – in Eurasia and North Africa. As far back as the ancient world, strongly urbanized states and territories had been formed. For instance, the towns/states of ancient Greece, and the whole territory of this region had been urbanized.

Table 6.3. Growth of the size of the urban population from the beginning of our era until 1995 (Fedorov, 1999).

	Years								
	1	800	1900	1950	1960	1970	1980	1990	1995
Urban population (billions of people)	0.029	0.081	0.124	0.138	1.033	1.053	1.752	2.277	2.584
Share of total population (%)	3.0	5.4	13.6	29.3	34.2	36.6	39.4	43.1	45.2

Urbanization acquired global scales in the 16th century after the great geographical discoveries, when the Europeans, carriers of agricultural technologies and urban culture, discovered new continents and began to develop them. However, this process was slow with towns appearing as the agricultural territories expanded. For instance, in the USA this process continued for about 200 years and, as a result, urban settlements appeared across the whole country. By the beginning of the 20th century, urbanization – the process of formation of urban settlements over the globe – was coming to an end. Most of the population still lived or were returning to rural localities. In the second half of the 20th century, however, the size of the urban population started growing rapidly once more, due initially to rural people moving to towns (Table 6.3).

If in the developed countries, which, as a rule, have a 1,000-year history of the formation of the urban settlements, the size of the urban population doubled between 1950 and 1990, in the developing countries during the same period it increased by a factor of 4.7. Here the number of towns of all sizes grew, as well as the number of large cities and their populations. In developed countries the population of some large cities even decreased (Table 6.4). In developing countries the poorest populations were rapidly moving from rural areas to towns.

Important features are observed in the changes of the number and dimensions of towns. The number and population of towns with population sizes from 100,000 and more grew rapidly: between 1950 and 1980 the number of such towns increased from 946 to 18,886, and the population increased by 677 million people. Nevertheless, a considerable part of the population moved to large cities with populations greater than 1 million. As a result, 30% of the urban population in Europe, 37% in Asia, 40% in Latin America, and 47% in North America live in cities containing 1 million people or more.

At present, the developed countries are most urbanized. In 1995, 74.9% of the population lived in these cities. In less developed countries, the urban citizens constituted 37.6%, and in the least developed countries 22.4%. However, among the developing countries there is a highly urbanized region – South America, where the urban population constitutes 78% of the country's population. In Asia and Africa only 34% of the entire population of these continents live in cities. The most urbanized regions are the rich oil-extracting countries of the Middle East, where urban people constitute 80% of the population. Large cities with populations of 1

Table 6.4. Distribution (given in %) of population in four categories of towns between 1950–1980 (*World Resources*, 1989).

Towns categorized by the size of population	Below 100,000	100,000–999,999	1,000,000–3,999,999	>4,000,000
Globally				
1980	38.1	27.9	18.2	15.8
1970	38.9	29.3	18.0	13.7
1960	40.8	29.6	16.1	13.4
1950	42.5	30.2	15.3	11.9
Developed countries				
1980	36.8	29.8	19.3	14.1
1970	37.1	30.3	18.3	14.2
1960	39.6	30.0	18.3	14.2
1950	42.3	29.4	15.7	12.5
Developing countries				
1980	39.1	26.3	17.3	17.3
1970	40.8	28.2	17.8	13.2
1960	42.4	29.2	15.9	12.5
1950	42.9	31.4	14.7	11.1

million or more have grown especially rapidly. In 1960 there were 114 such cities, 62 of them being found in developed countries and 52 in developing countries. By 1980 these cities already numbered 232, of which 113 were in developed countries and 119 in developing countries. By the beginning of the 21st century such cities numbered 321, with most of them assigned to developing countries. In 1950 only one very large city (with a population of more than 5 million) was to be found in developing countries – Buenos Aires in Argentina. By 1985, of the 25 largest cities of the world 20 were located in developing countries. By the beginning of the 21st century there were 14 "city states" with populations of 10 million or more (*Review of the Global Economic and Social Situation ...*, 1996).

Sources of growth of the urban population

The growth of the urban population took place for two main reasons: natural population increase and migration of people from rural regions. From 1950 to 1990 the number of urban people in developed countries grew by 0.8% every year, on average, which was close to the natural growth of the population, whereas in the developing countries this growth constituted 3.6%, which is much higher than the natural population increase. In the underdeveloped countries the natural population increase prevailed, especially between 1960–1970. During these years the growth of the urban population due to migration from rural regions constituted 40–44%, but in the 1980s it reached 54.3%, mainly due to the cities of China and Indonesia, where social–political changes took place, opening the way for rural people to move to the city (Fedorov, 1999).

Urbanization has affected masses of people. During the 20th century (until 1995) the size of the urban population increased by 2.36 billion people. During this time the global population increased by 4 billion. Hence, the urban population contributed more than half of this population increase. Of them, about a half were born in cities, the other half came to cities from rural regions. One in every four people of the growing population became a city immigrant. This enormous flow of people to the cities from rural regions is the greatest social and ecological phenomenon of the 20th century. Countryfolk moved to the cities in the hope of obtaining jobs, better living conditions, an education, and a profession. However, in developing countries many settlers from rural localities found life in the cities fell short of their expectations, since there were no real employment policies in the cities, causing growth in both unemployment and partial employment. This resulted in a crisis within cities, expansion of the slums, homelessness, and other negative social phenomena. This crisis was also determined by the demographic situation, since a high birth rate and a rapid increase of population meant that poor countries could not provide places for work in rural areas or cities. In addition, a considerable amount of jobs were concentrated in the non-productive and low-productive branches of industry.

Over recent years, the problem of unemployment has appeared in developed countries as a consequence of a low rate of economic growth, the possibility of it dropping further, and active propagation of new, especially information, technologies. The level of unemployment is especially high in the developed countries of Europe where it varies between 6–14%.

The share of urban population and the rate of urbanization characterizing global regions is given in Table 6.5.

It follows from Table 6.5 that urbanization in developed countries is very high. Three-quarters of the population live in cities. Nowadays, the rate of urbanization is decreasing. At the same time, the developing countries (less urbanized, with a little more than one-third of the population living in cities) have increasing urban populations – 4 times higher than in developed countries. However, the developing countries are rather inhomogeneous. There are three large regions among them: Latin America, Africa, and Asia. Latin America and the Caribbean Basin are highly urbanized. For 20 years (from 1975 to 1995), Latin America not only caught up with the developed countries in its extent of urbanization but even surpassed them. In Argentina, Chili, Uruguay, and Venezuela, more than 80% of the population live in cities. A specific feature of this region is a very high population density in one city. More than one-quarter of the urban population live in the capitals of Argentina, Bolivia, Dominican Republic, Costa Rica, Cuba, Panama, Peru, Puerto Rico, Uruguay, and Chile. Urbanization of this region was prompted

Table 6.5. The share of the urban population and the rate of the growth of the urban population for continents, regions, and individual countries, % (*Review of the Global Economic and Social Situation*, 1996).

Global region	Share of the urban population		Rate of the urban population growth for 1975–1995
	1975	1995	
Developed countries	69.8	74.9	0.4
Developing countries	26.7	37.6	1.7
Underdeveloped countries	14.2	22.4	2.3
Global total	*37.7*	*45.2*	*0.9*
Eastern Africa	12.3	21.7	2.8
Central Africa	26.6	33.2	1.1
Northern Africa	38.6	45.9	0.9
Southern Africa	44.1	48.1	0.4
Western Africa	22.6	36.6	2.4
African continent	*25.2*	*34.3*	*1.6*
Eastern Asia	25.1	36.9	1.9
China	17.3	30.2	2.8
Japan	75.7	77.6	0.1
South of Central Asia	22.2	28.8	1.3
South-East Asia	22.3	33.7	2.1
Western Asia	48.4	66.4	1.6
Asian continent	*24.6*	*34.6*	*1.7*
Eastern Europe	59.9	70.4	0.8
Northern Europe	81.7	83.7	0.1
Southern Europe	59.1	65.1	0.5
Western Europe	77.8	80.5	0.2
European continent	*67.1*	*73.6*	*0.5*
Caribbean Basin	51.0	62.4	1.0
Central America	57.1	68.0	0.9
South America	64.2	78.0	1.0
North America	73.8	76.3	0.2
Americas	*61.3*	*74.2*	*1.0*
Oceania	71.8	70.3	−0.1
Australia – New Zealand	85.3	84.9	0.0

by industrialization aimed at substituting imports, and by a high level of birth rates in the cities in the 1950–1960s.

Africa remains, so far, a low-urbanized continent. Here a little more than one-third of all people live in the cities. The rate of urbanization here is very high. Now the African cities are similar to the cities of South America in the early second-half of the 20th century. The annual rate of urban population growth, constituting 4.4%, is determined by the high natural population increase and by the movement of country folk to the cities. In contrast to Latin America, the urban population in Africa is concentrated, as a rule, in towns with populations of 500,000 or less. So far, the population of only two cities on the African continent exceeds 5 million: Cairo (Egypt) and Lagos (Nigeria).

In 1995, 3.3 billion people lived in Asian developing countries (58% of the global population, including 46% of the global urban population). Here can be found the most urbanized countries of the world – Bahrain, Hong Kong, Israel, Qatar, Kuwait, and Singapore, in which more than 90% of the population live in the urban regions. At the same time, countries with the lowest level of urbanization are: Bhutan, Eastern Timor, and Nepal, where the share of the urban population constitutes less than 14%. Finally, there are five countries in Asia with the largest populations: China, India, Indonesia, Pakistan, and Bangladesh, with relatively low levels of urbanization over the whole region. In 1975–1995 the developing countries in Asia were characterized by a high rate of urban population growth – 3.9% yr^{-1} (*Review of the Global Economic and Social Situation*, 1996).

In the middle of the second-half of the 20th century the population in urban areas of developed countries decreased, caused by suburbanization. It was a process of reducing the size of the population of large cities via the return of people to the surrounding smaller urban settlements. In the USA, a large-scale transition to suburbanization was observed in the late 1960s and early 1970s. Along with a departure of people from cities smaller than a megalopolis (0.3 million people), the population of small towns increased by 0.4 million people per year.

Studies carried out in the 1970s showed that similar processes took place in European countries and in Australia. In this connection, a scenario for further development of this process for the 1980s was suggested. However, the situation proved contrary. The latest data showing that, again, the population was concentrated in large cities, as observed in the USA. The rate of population growth in large cities, for instance, in Paris and London, has increased over the last years (*Review of the Global Economic and Social Situation*, 1995).

Suburbanization in the 1970s showed that the main cause of changes of inner migration flows was the changing structure of the economy, which made it possible to distribute enterprises outside large industrial centres, which was determined by the appearance of science-driven technologies. A change in the views held by the public concerning where to live also affected population distribution. Suburbanization prompted governments to invest in infrastructures, including transport, health services in small towns, and to pursue a policy of decentralization and building of new towns. However, the economic recession in the late 1970s and early 1980s reduced investment opportunities and caused a cessation of many social programmes, which again led to the concentration of populations in large cities. Nevertheless, this experience showed the possibilities of regulating the growth of large cities.

A most important process determining the growth of the number of cities is the transformation of rural settlements into the urban centres. This transformation results from production development not connected with agriculture, as well as from the growth of trade and development of infrastructure and administration.

Table 6.6. Growth of the urban population in Russia and the USSR and change of the share of the urban population with respect to total population between 1851 and 1939.

	Year						
	1851	1863	1897	1914	1917	1926	1939
Population (thousands of people)	3482	8157	16,785	24,700	26,300	26,314	55,910
Share of total population (%)	7.8	10.6	13.0	15.6	15.6	17.9	32.8

Table 6.7. Growth of the population of Moscow and St Petersburg in the period between 1770–1939 (figures in thousands of people).

	Year							
	1770	1780	1870	1897	1900	1917	1926	1939
Population of St Petersburg	158.8	220.2	628.3	1265	1418	2165	1690	3191

	Year							
	1785	1860	1885	1897	1910	1917	1927	1939
Population of Moscow	180	360	800	1038.6	1500	1701	2029	4137

To a great extent these processes determined the trend of urbanization. In the international practice, an increase of urban population due to transformation of rural settlements into urban is called re-classification, and usually this is combined with the increases caused by population migration.

The process of reclassification is characteristic of the developing countries since in the developed countries a stable enough relationship was fixed between the urban and rural settlements. Despite the growth of secondary and small towns, the share of their population with respect to the total urban population in the developing countries decreased due to a very rapid growth of cities with populations of 1 million or more. In the developed countries the share of people in towns with populations below 100,000 also decreased and in cities with populations of 100,000–1 million remained stable. In the cities with populations of 1–4 million it grew and in larger cities it was stable.

The development of cities in the territory of Russia had been taking place for more than 1,000 years. In 1650, there were 226 towns in the developed European part of the Russian territory. In the 17th century, 59 new towns appeared, some of them being in Siberia (Tomsk, Irkutsk, Yakutsk, etc.). In the 18th century, 206 new towns were added. Data listed in Tables 6.6 and 6.7 (*Great Soviet Encyclopaedia*, vol. 12, 2nd edn) are tied to certain years close to important events and population census. In particular, Table 6.6 demonstrates a standstill in the growth of cities and the urban population in the period between the October revolution and the beginning of a new economic policy (1926). In some cities, after 1917 the population decreased sharply

Table 6.8. Dynamics of the growth of the urban population in the USSR and in Russia between 1959 and 1989.

Dynamics of the urban population	Year			
	1959 (USSR)	1970 (USSR)	1975 (USSR)	1989 (Russia)
Urban population (millions of people)	100	136	1531	108,419
Ratio of urban to total population (%)	48	56	62	73.4

in connection with the Civil War and the first wave of emigration from Russia. At the same time, in the period between 1926–1939 (a period of industrialization), the urban population started rapidly increasing. This increase was determined, first of all, by migration of the rural population. The migratory contingent constituted 18.5 million people (i.e., more than 60% of the increase). Reclassification (i.e., conferring the status of towns to rural settlements) added 5.8 million to the increase. The natural increase provided only 17.9% of the total growth of the urban population. Thus it took only 13 years to double the size of the urban population in the USSR, whereas in the USA it took 30 years, and in the UK 70 years.

The period of industrialization became a period of great migration for peasants to the cities, and for wealthy peasants to Siberia, to exile. Migration of peasants to cities was not always voluntary. Some were driven by hunger while others were mobilized to build the objects of socialistic industry. Naturally, the new urban population was not provided with homes. As a result of the concentration of dwellers, the so-called "communal flats" appeared. On the outskirts of cities a chaotic uncontrolled building of dwellings began, which favoured an appearance of slums. The development of the municipal and transport infrastructure lagged drastically behind. This negative legacy is still preserved in many cities. Nevertheless, the rapid urbanization made great masses of the population familiar with the urban culture, hygiene, medical aid, as well as improved access to education and the arts.

Destruction and loss of population during the Second World War slowed down the process of the growth of cities, but, nevertheless, by the year 1959 the size of the population in cities approached almost half of the USSR population. The share of the urban population in Russia (from the data of the 1998 census) reached almost three-quarters of the population of the country (Table 6.8) (*Great Soviet Encyclopaedia*, 2nd edn; Losev et al., 1993).

Thus in the 20th century a "quiet" revolution took place, during the lifetime of less than two generations a great proportion of the global population drastically changed from rural to urban lifestyles. This has led to great socioeconomic changes, changes in human behaviour, and an acceleration of environmental degradation.

6.4   GLOBALIZATION

Human beings became a planetary-scale phenomenon tens of thousand years ago when small groups of people populated land suitable for life (Antarctica remained unpopulated). Thousands of nations and peoples had been formed, with languages and cultures of their own, from more or less homogeneous groups of people with a single material culture.

Linguistic ability is inherent in humans, and language is the product of accumulated culture and social structure. People cannot exist on their own without elements of altruism and mutual assistance. Life for them is only possible in a complex social structure. The resulting populations of ethnic communities prevent complicated interactions in the social structure from disintegrating or resorting to competitive interaction. Forms of such an interaction depend on the level of culture. As long as humans – the cultural state of the biosphere – exist, nations will compete. Wars have been the form of this competition during the whole period of the civilization existence. Only towards the end of the 20th century did military clashes begin ceasing, with the transition to economic competition, which revealed a political system supposedly most competitive – a democratic civil society developing on the basis of scientific–technical progress and market economy. It seems that it is the competitive interaction of the ethnic community that serves as a stimulus for globalization or rather global changes, this being not only an association and interdependence but also a division and stratification in various aspects – from linguistic and cultural to socioeconomic and governmental. From the historical point of view, it is especially apparent that many modern "signs" of globalization existed long ago, and present generations, with an access to a huge flow of diverse actual information, consider them the signs of our modern times.

One of the convictive signs of globalization is the present-day economy. However, in fact, the first economic global integration had taken place after the great geographic discoveries and worldwide propagation of agricultural technologies. The second wave of global integration resulted from the Industrial Revolution, when the epoch of "free trade" began along with a sharp growth in the export of goods and raw materials, as well as industrialization and modernization of the world community. Industry of the developed countries was reorganized to process the imported raw materials and to carry out a global-scale selling of produce. Although between 1850 and 1914 the population increased only by a factor of 1.5, the product exchange in international trade increased 10 times. Foreign investments grew rapidly, too: from the beginning of the 20th century until the First World War global production increased by 40% and the physical turnover of the global trade by 60%. During the same period the volume of foreign investments doubled. In the 19th/early-20th centuries the world colonial empires promoted globalization – world trade was based on the colonial status of the territories where one-third of the global population lived. In the same period, the labour resources moved mainly from Europe to America, Australia, southern Africa, and Siberia. Up to 20% of the annual increase of the population left Europe in search of a better lot in other countries. From 1850 until 1900 the population in Europe increased from 250 to 450 million. International transport was rapidly developed and new types of communication appeared: large ships and dirigibles were built as well as the Suez and Panama Canals. Telegraph cable was laid at the bottom of the Atlantic (1858), telephone and radio appeared, and the international mail service intensified.

There were other signs of what is called "globalization": the First International Peace Conference was convened (1899), the international Nobel prize was established (1901), and the language of international intercourse – Esperanto – was invented (1887). Most of the global population lived in the countries with convertible currency, the frontiers were open – passports were not needed to cross them (passports were only used in Russia and the Ottoman Empire at that time) to find work in a foreign country. The first Olympic games were organized. Finally, notions appeared such as "imperialism" and the slogan "workers of the world, unite". It is interesting that England at that time was not only the leading industrial power but also the basic supplier of fossil fuel to the world – coal, which it exported to half the countries in the world.

However, competition increased and continued up to the First World War, when the process of globalization stopped (at least for 50 years). The war itself destroyed the international trade system and labour division and separated the economies of countries. After the war, cardinal changes took place aimed at closing the process of globalization: international trade was sharply reduced and countries reinforced their custom frontiers to protect the post-war industry branches from market competition. Export of finances was reduced (as a % of the gross domestic product). Migration of labour power decreased, many countries imposed restrictions on entry and on permission to get work for foreigners. For trips abroad a system of passports was introduced. Many countries pursued a policy of developing the import-substituting industries.

The 1929–1930 economic crisis further reduced international contacts. During these years international trade decreased by two-thirds. The greatest producer in the world – the USA – isolated itself in the economic space of its own, believing that it could do without foreign trade. The same took place in fascist Germany. In the USSR export decreased by two-thirds. All this testified to an aggravation of competition and suspense of the associated problems during the First World War. Revenge seeking and nationalistic spirits in Germany and a mass ideological impact on the German people played an important role, too.

After the Second World War the processes of globalization were very slow. The consequences of this war showed that such forms of competition had become obsolete for three reasons:

1   it became clear that these forms of competition were very unprofitable economically, since the economic basis of most of the countries that participated in the war were almost completely ruined;
2   nuclear weapons had appeared, which in case of war could destroy not only economies but also every living thing on the planet; and
3   the war led to the formation of superpowers.

These superpowers were the USA and USSR, which had approximately equal offensive/defensive potentials. They competed in not only the sphere of nuclear weapons but also in economic and social spheres, though this competition was not apparent. These were specific features of the beginning of the process of "new globalization", in which the confrontation of the powers played a substantial role in the development of integration within the spheres of influence of each of them.

As late as the 1960s, globalization in the economy was not discussed, and its perspectives were estimated rather pessimistically. But from the mid-1970s the indicators of global trade, export of long-term capital, and volumes of foreign investments with respect to GDP reached the same relative estimates as at the beginning of the century. The scientific–technical progress – transition from steam engines to electric energy and internal combustion engine, and the wide use of oil and gas – played an important role. However, in the late-20th century, the process of globalization in the economy differed substantially from that in the late-19th/early-20th centuries. The last century became a period of the formation and growth of transnational corporations without national frontiers and with a free capital flow. The two hundred largest transnational corporations of the world ensure 40% of the volume of sales in the car industry and oil extraction and processing. At present, international financial organizations have been formed – the International Monetary Fund and the World Bank. Modern globalization has been followed by a rapid growth of financial markets (currency, credit, and stock), which are isolated from the sphere of production and trade. The finance capital is capable of speculative self-expansion and can substantially affect the national financial systems. A substantial difference of modern globalization is also a creation of regional associations of countries, one of which is the European Community, where the states decided to reject some attributes of sovereignty. This regionalization of the world is considered as a step towards globalization, but in fact it is a new form of competition both within regions and between a region and its immediate surroundings.

One further reality of the new globalization is an appearance and development of global problems affecting all or most of the countries. The ecological crisis is one of the most important problems.

In the late-20th century, other processes appeared: the transition of the world to a uniform material culture, whose symbols became the car, TV, cell telephone, and Kalashnikov machine gun; contiguity of the standards of higher education; and unprecedented development of the international communication network (Internet) and computer technologies.

At the same time, as in the past, it is not only the integrating processes connected with globalization that take place, there are also the separating processes, parts of which are generated by globalization itself. The apparent (often forced) division of countries into First, Second, Third, and even Fourth World nations has taken place. The gap between the rich and the poor widens: during the second-half of the 20th century the ratio between the incomes of the "golden" billion and "poor" billions increased from 13 : 1 in 1960 to 60 : 1 at the time of writing, though during this period the total volume of consumption increased six-fold. The number of countries on the planet increased almost three-fold. Nationalism, fundamentalism, and fanaticism are propagating, and on their basis so is global-scale terrorism. The world in different places lives by different time coordinates – from feudalism and "wild" capitalism to democratic societies. But most important is the fact that in the 20th century, civilization confronted nature, and now it is clear that the future development of our civilization is limited by ecological conditions.

Processes of globalization, hopefully, seem to be a mechanism to form a new regime. Nevertheless, historic experience shows that it is a reversible process. If the ecological conditions are not taken into account and the physical growth of civilization is not limited, a global chaos can occur instead of global order.

6.5   GLOBAL SOCIAL–ECOLOGICAL CRISIS AS A RESULT OF GROWTH IN ALL DIRECTIONS

During the last decades, discussions of the problems of the ecological crisis went beyond the framework of scientific discussions within the world scientific community. Concern for the fate of the biosphere has involved masses of people, triggered "green" movements, and led to the formation of political parties. It affected the states – their governments created organizations for nature and environment protection – as well as respective environmental infrastructures. Also, international and non-governmental organizations were formed to protect the environment and nature. Finally, nature protection legislation appeared and respective international agreements were completed. On the basis of nature protection technologies, business created markets with a turnover of several billion. All this is an apparent confirmation of the global ecological crisis on our planet, which is a result of the exponential growth of the global economy and population (i.e., growth in all directions). Anthropogenic impacts have acquired a planetary scale, and the whole energy production of the present civilization will end up destroying life-supporting natural systems.

Growth in all directions in the 20th century

The global population increased from 1.6 billion in 1900 to 6.08 billion people in 2000. During the first-half of the century the increase constituted about 1 billion, and during the second-half – more than 3.5 billion, despite a reduction in the increase from 2.2% in 1964 to 1.4% in 1997. The gross global product (GGP) increased from US$3.8 trillion dollars in 1900 to US$43 trillion (US dollars of 1997). The main increase fell in the second-half of the century, when the GGP increased from US$6.4 to 43 trillion. On the whole, the GGP increased 18-fold and the per-capita increase constituted from 1900 to more than US$7,000 (i.e., by a factor of 3.5). The area of arable lands doubled – from 700 to 1,500 Mha and the area of irrigated lands increased six-fold. The grain yield increased from 400 to 1,900 $Mt\,yr^{-1}$ (i.e., more than 4.5 times). The same increase was observed from the moment of an appearance of agriculture until the 20th century, but the per-capita grain production increased by only 60 kg – from 250 to

310 kg. The global energy consumption in tons of oil equivalent increased from 911 Mt in 1900 to 10,000 Mt in 2000. During this period, water consumption increased from 400 to 4,000 km^3 per year (Danilov-Danilyan and Losev, 2000; *State of the World 1999*, 2000).

The global account of materials consumption was only started in the 1960s. From this time to the end of the century the consumption of wood (not as fuel) increased 21-fold, of synthetic materials (on the basis of fossil fuel) 56-fold, and on the whole, based on all materials 24-fold (*State of the World 1999*, 2000). Anthropogenic flows of many materials exceeded the natural flows by a factor of 2 or more. Nevertheless, the enormous technical progress has not solved global social problems. Hunger and poverty are still commonplace. Moreover, in the late 20th century the numbers of poor increased (Table 6.9).

It follows from Table 6.9 with an addition for values for the year 2000, that the absolute amount and share of the poor have increased in four of the six regions. In one region they have not changed, and only in Southern Asia have they decreased markedly. On the whole, the global share of the poor by the year 1990 decreased negligibly. During the last decade the amount of poor, once again, started increasing and from 1990 to 2000 their numbers increased by 367 million. During the same period the global population increased by 800 million. Since the population increment took place mainly due to the developing countries, a conclusion can be drawn that 46% of this increase are doomed to starve and/or live in poverty. In the world, 1,500 million people live on US$1 dollar per day or less and 2 billion people live on US$1–2 per day. In the developed countries the amount of poor has also grown, as determined by the standards of these countries. So, in the USA the lowest percentage and amount of poor was observed in 1973 – less than 11% or 23 million people. In 1999 the share of the poor in the USA constituted 11.8% or 32.3 million people (Dalaker and Proctor, 2000). Half the poor in the world live in villages, the other half in cities, where they face a huge amount of social problems.

Table 6.9. Changes of the numbers of poor in developing countries between 1985 and 2000.
From *Development and Environment* (1992) and *Advancing Sustainable Development* (1997)

Region	Numbers of poor (millions)				The poor (%)	
	1985	1990	1996	2000	1985	1990
Southern Asia	532	562	–	–	51.8	49.0
Eastern Asia	182	169	–	–	13.2	11.3
Sub-Saharan Africa	184	216	–	–	47.6	47.8
Middle East and Northern Africa	60	73	–	–	30.6	33.1
Latin America and Caribbean Basin	87	108	–	–	22.4	22.5
Eastern Europe (without USSR)	5	5	–	–	7.1	7.1
Total	*1,051*	*1,133*	*1,300*	*1,500*	*30.5*	*29.7*

The social problems of the cities

As far back as 400 BC, Plato, who lived in the most urbanized state of antiquity, wrote that any town, small as it may be, consists of two towns – a town for the rich and a town for the poor. Nowadays, 1.1 billion urban people suffer from a highly polluted urban environment, 220 million suffer the deficit of fresh water, 420 million have no access to canalization, and 600 million are homeless. During the 20th century the size of the urban population increased 15-fold, and in the near future it is estimated that more than half the population of the world will live in cities (Fedorov, 1999; *State of the World 1999*, 2000).

Poverty is not just about a shortage of food, it is also represented by a lack of choice, the impossibility of getting an education even in primary schools (now 140 million children have no possibility to study), an aimless life, limited chances for health care, unattainability of medicines and medical aid, a lack of prospects, and a general struggle for survival. The conclusion is obvious: a global social crisis has not only taken place it has grown during the last decades of the 20th century (Table 6.9).

The social crisis is unavoidably linked with the ecological crisis. They are mutually intensifying. The latter is developing on a global scale and covers the whole geographical environment (biosphere). This is confirmed by numerous observational data, which definitely point to unidirectional changes of the concentration of basic nutrients and other substances in the atmosphere, surface waters, and in the soil, as well as to a rapid reduction of biodiversity. But most serious violations are connected with the disappearance of natural ecosystems on land as a result of human activity and technologies. Humans have practically destroyed the ecosystems that once covered 63% of the land areas of the Earth. Major parts of these ecosystems were destroyed in the second-half of the 20th century. *The ecological results of the development of the terrestrial technological civilization turned out to be devastating: the ecological indicators of the quality of human life and health are getting worse. The inert technological development of civilization is going on, based not on the scientific achievements but on experience of the past, within the framework of the modernistic stereotype.*

Ecological results of the development in the 20th century

The area of land territories with destroyed natural ecosystems increased from 20% in 1900 to 63% in 2000. The per-capita area of arable lands under cereals decreased from 0.22 to 0.12 ha, and in the countries where half the population of the world live (China, India, Indonesia, Pakistan, Bangladesh and Nigeria) – from 0.52–0.16 ha at the beginning of the century to 0.07–0.03 ha at the end of the century. In the 20th century the renewable resources (air, water, soils, vegetation, and animals) ceased to renew in the same amount and quality within their natural fluctuations. Biodiversity is rapidly reduced: of 242,000 studied species of plants, 33,000 species are under threat.

By the end of 2000, man learned to synthesize 18 million substances, most of them being absent from global biogeochemical cycles. About 200,000 substances circulate in the market. The impact of most of them on man and other organisms is unknown. In the atmosphere of industrial cities one can discover hundreds of pollutants, and in human tissues they number almost 2,000.

During the last 25 years, 29 new diseases have appeared, including dangerous ones like Lyme disease, Ebola, Legionnaires' disease, AIDS, etc. (Colborn et al. 1996; *State of the World 1999*, 2000; Losev, 2001b).

7

The scientific basis of life sustainability

7.1 FORMATION OF THE ECONOMIC AND ECOLOGICAL STEREOTYPES

Management in the form of war and destruction

A brief review has been made in previous chapters of the development of civilization. During this period man constantly attacked nature as well as the transforming the geographic environment to suit his needs. Such "management" of the environment was of primary importance at all stages of civilization development. In fact, it was an expansion, and, as a result, man developed new territories and tried to convert them to a state optimal for himself. Optimization was mainly determined by economies aimed at producing a greater volume of diverse products and hence at obtaining profit. For this purpose, forests were felled, marshes were drained, steppes and savannas were destroyed, and empty territories transformed into arable lands, pastures, urban and industrial territories, roads, and other means of communication. It was a war with nature, in which human beings believed they could not lose, since the rate of progress (i.e., the rate of creating new technologies) continued to grow. In the late 20th century it exceeded the rate of evolution (the rate of formation of new species of organisms) by at least five orders of magnitude. Before the last century, all destruction of the environment had taken place within the carrying capacity of the Earth (see Figure 3.1), when the regulating and stabilizing mechanisms of natural ecosystems continued functioning, and the carrying capacity of the biosphere was never overstepped. Quite recently, the economic activity of humankind has overstepped such a limit, with the growth and acceleration of economic development in all directions (see Chapter 6). In the process of this frantic activity, humankind has overstepped and rejected a basic ecological restriction – the limits of the carrying capacity of the global ecosystem.

If during the earlier centuries human communities on some territories exceeded the limits of the carrying or economic capacity of ecosystems, this was accompanied by food crisis and disintegration of state (see Section 4.3). At the beginning of the 20th century, humankind exceeded the limits of the carrying capacity of the global ecosystem. This is verified by observational data: significant changes in the nutrient concentration, as well as of other elements and substances in the environment. As a result, anthropogenic flows of some nutrients exceeded natural flows (Vitousek et al., 1997), and, on the whole, the growing trend of the concentration of these substances exceeded their natural fluctuations.

The stunning progress of so-called "civilization development", especially in the second-half of the 20th century, and the rapid growth in computer technologies as well as the World Wide Web have become a symbol of humankind's abilities. Humankind holds enormous power in its arsenal of nuclear weapons – enough to kill all living beings on the Earth. At the same time, not a single technology has been developed so far in any sector of the material sphere of human activity that has been designed to tackle pollution. The chaotic activity that is civilization development has no strategy, apart from the desire to grow in all directions. So, humankind continues to live in accordance with stereotypes formed over centuries (Kondratyev and Losev, 2002).

During the 10,000 years of civilization existence, humankind has changed its way of life in the world only twice. The first change took place during the transition of humans from hunter–gatherers to growers of domesticated plants and animals. During this transition, humans rejected zoomorphic gods, stopped deifying the forces of nature, and created "humanized" gods, who were then combined into a single God standing at the centre of the world. Relatively recently, at the beginning of the Renaissance, humans placed themselves at the centre of the world, and God was returned to the Heavens. A period began in which knowledge, science, technologies, the free market, and rights and liberties of humans marginalized the earlier ideas of the world. The modernization of society has led to an accelerated change in the world during the last 300 years (especially during the last century): economies and populations grew exponentially, colonial empires disintegrated, and many new states appeared. An experiment, investigating the centrally managed modernization of society, revealed a reduced economic efficiency compared with the market system and a greater efficiency in destroying the environment. Consumption of renewable and non-renewable resources grew especially rapidly (in particular, water and fossil fuels).

Before the 20th century, modernization took place when the environment was sustainable, with abundant resources, relatively limited local violations of natural ecosystems, and a limited pollution of the environment. At the beginning of the 20th century, global changes in the environment began, although before the 1960s they were not apparent because of preoccupation with two World Wars, the Great Depression and the period of reconstruction. Also, at this time there was no reliable system to monitor the environment. Only in the 1960s did the world community "discover" the environment, and new observational tools began to monitor the growing rate of changes in environmental characteristics. The latter has led to economic and social outlay and to violation of human health. The global ecological crisis was increasing. As a result, national and global nature–protection infrastructures were formed in the 1970s. But humankind still has not realized that global changes of the environment mean a civilization–nature clash, and that these changes confront life materialized as biota – other organisms numbering 10^{28} combined into ten million species. These changes meant that civilization was at war with other living beings in the interests of only one species – *Homo sapiens* – and a very narrow circle of cultural plants and domestic animals.

The growing proportion of organic material produced by photosynthesis is directed into an anthropogenic channel of consumption. People destroy more and more areas of natural ecosystems (landscapes) – the habitats of various organisms.

Changes of natural landscapes over the last 300 years

The global area of forests and woodland decreased by 12 million km^2 (19%) with meadows and other grass landscapes reduced by 5.6 million km^2 (8%, with many meadows transformed into pastures). Pastures widened from the year 1700 from 4–5 million km^2 to 31–33 million km^2 at the expense of meadows and steppes. The area of arable lands increased by 12 million km^2 (400%). In 1700, arable land covered 3–4 million km^2, and in 1990 reached 15–18 million km^2, mainly due to deforestation.

Except for a rather small circle of scientists, nobody else has posited the question of the necessity to study the consequences of such destruction, the changes of views on nature, and the study of the functions of ecosystems and the laws of such a system – biota and its environment – within which and due to which humankind has survived. However, for 300 years of modernization and huge achievements, humankind has developed a system of stereotypes, which still does not allow one to realize the necessity of a thorough study of the laws of the interaction of biota with its environment (biosphere or geographical environment) and the resulting limitations of the unrestrained growth of civilization. So far, they have been studied separately: landscape scientists studied the abiotic part of the geographical environment and biologists its living components (see Chapter 3). Even ecology, as Rozenberg (1999) noted, is still understood to be either a section of biology or the science of the interaction between human society and the environment, including biota. Today, it is clear that in the so-called biological ecology the large-scale and strong impact of human economic activity on biota and its environment should be studied. Therefore, it is necessary to have a clear scientifically-grounded idea of the functions of natural ecosystems and the consequences of their destruction by man.

Brown and Flavin (*State of the World 1999*, 2000) of the Worldwatch Institute wrote in this regard: "Will we recognize that the world is naturally limited, and reconstruct our economy respectively or will we, as before, aggravate the ecological consequences of our progress until it turns out that it is too late to turn back? Aren't we concentrated on the world in which the accelerating changes overtake our ability to control the economy, undermine our political institutions and lead to a large-scale destruction of ecosystems, on which our economy depends?" The latter question can be answered positively, if humankind acts in accordance with the existing stereotypes of modernization, but with one very important correction: that not only do economies depend on the destruction of ecosystems but also the existence of humankind as a species and probably a considerable number of other species.

During most of their lives, humans are guided by formed and inspired stereotypes. This takes place in all spheres of activity, from bringing up children to science and the economy. During modernization, the economy formed stereotypes of its own, and during the second-half of the 20th century, ecology created stereotypes of its own. They were necessary and correct until a certain time. After the global-scale confrontation of civilization with nature these stereotypes have become a kind of incorrect information, since this confrontation, in fact, has changed the world completely. We will dwell upon some stereotypes.

There is a well-known statement that after the agricultural (Neolithic) revolution humans moved on from the "appropriating" economy to a "production" economy. In fact, a transition took place from the appropriating to hyperappropriating economy: just like a primitive human (free-of-charge and heartlessly), recent humankind takes for itself (as only one of the 10 million biological species on Earth) the Earth's renewable resources and does it on global scales. Scientific–technical progress ensures an increasingly efficient means of extracting resources. While extracting some resources, humans simultaneously destroy natural eco-systems, completely or partially transporting huge masses of soil and grounds (on average, up to $50\,t\,yr^{-1}$ per capita). Some technologies are water-based, consuming water at the rate of about $800\,t\,yr^{-1}$ per capita (Losev et al., 2001b).

What is the end-product of the production economy? It is believed to produce products, but this is not true. In fact, the enormous system of the world economy produces nothing but waste, which is divided into production waste and consumption waste. Admittedly, without leaving behind waste humankind would never have known its history, and archaeology would not exist as a science. Consumption waste is divided into short-lived (products used over a single year) middle-lived (used during a period up to 10 years), and long-lived (those used for more than 10 years).

The whole world struggles against waste, trying to resolve this problem. It is however, impossible to struggle against the law of conservation. By this law, waste can only be buried or transformed into another phase (e.g., from solid into gas or from toxic into less toxic). It can be used to produce another product, which again will become waste. However, the latter process cannot be endless and from the point of view of energy used is usually unprofitable. Every year the global economic system deals with 300 Gt of substances and 13 TW of power are spent processing them.

In the world such notions as "zero-waste technologies" and "ecologically pure technology" are widely used. However, there are no such technologies in the sphere of material production. Even in the tertiary economy (service economy) there are material flows and waste and destruction of natural ecosystems. To build a bank, supermarket, hospital, etc., a territory is needed as well as metal and many other building materials. In addition, constant provision with heat or cold (or both), electric energy, and water is also required. In such building processes waste is often produced and often very dangerous (i.e., medical waste in hospitals).

The notion of "renewable resources" has passed away. Freshwater and atmospheric air are no more renewable both in quality and in quantity, since changes of concentrations of their constituents exceed their natural oscillations, and new substances appear, as well as an additional mass of suspensions appearing in the land surface waters. Soils, vegetation, and animals are rapidly reduced both in quantity and in quality. The main mass of river water is polluted with organic and mineral substances.

An assessment of renewable resources by stocks are a settled stereotype in the resource sphere. In fact, they should be assessed by their flows. Only part of the flow may be used, otherwise a degradation and exhaustion of the resource begins as well as changes of other renewable resource components (i.e., there is a limit to their use), since all of them are closely connected (the "dominoes" principle). For instance, the whole annual increase of wood in the forest cannot be considered to be a wood stock, which can be taken from the forest ecosystem. Removal of such masses of wood will break the natural balance of the nutrient cycle and, hence, the stability of the forest ecosystem. When such a limit is exceeded, the process of shrinking forest areas begins, desertification sets in, and problems of emission of gases into the atmosphere and soil erosion arise. Tinbergen and Haavelmo (1991), Nobel prize winners in the sphere of the economy, justly note that the renewable resources are not simply resources but the life fundation. By using them beyond an admissible limit, humankind destroys the foundation of its life.

The present society is called industrial, and some countries are considered the countries of the information society. At the same time, like before, the existence of the present civilization is based on food production (i.e., agriculture) without which humankind cannot exist. A human is not an autotrophic being. Therefore, soils and agrarian activity, but not industry or information technologies (which can be considered only as the service systems), are basic values for humankind. The agrarian sphere of activity makes the existence of humankind possible. Industry for the agrarian activity is only a system to improve efficiency and increase productivity. Industry itself is a system to create a means to improve the comfort of human life. At the same time, industry is a source of risk. One should remember that in industrial enterprises, in air, water, and soil environments, in food products and in dwellings about 200,000 substances produced by industry on large scales, are in a permanent cycle. For 90% of these substances, their impact on living organisms, including humans, is unknown.

Soil is a slowly renewing but rapidly destroyed resource. Therefore, it is likely to be a shortage of food, not energy, that may lead to the first global resource crisis. Industry plays the role of an active destroyer of the soil cover (since for its development additional space is constantly required) and as a pollutant of large agricultural areas.

Energy production has stereotypes of its own

All energy used by humankind is eventually spent on the destruction of the natural environment and production of waste in the name of human comfort, or as it is sometimes called "civilization's armor". Objective observational data demonstrate huge changes caused by man due to the use of energy (mainly fossil fuel energy) in the environment. Fossil fuel provides 99% of the energy needed for industrial countries and 75% of the global requirement (*State of the World 1999*, 2000; Kondratyev et al., 2002a, b, 2003). The use of fossil fuel energy has resulted in the depletion of natural ecosystems for 63% of the land surface. Fossil fuel burning affects the Earth's climate and promotes emissions of enormous masses of dangerous pollutants into the environment.

The energy industry contributes significantly to the global-scale destruction of the environment. The threshold of energy consumption by civilization has already been exceeded (Gorshkov, 1995; Kondratyev, 1998; 1999a). In these conditions all the so-called alternative energy sources, which, for reasons unknown, people consider "ecologically pure", are also ecologically dangerous to the environment. At the same time, if the use of energy does not exceed a permissible threshold, any sources of energy are considered acceptable. In the conditions of the nature–civilization confrontation the problems of energy use should not be considered one of concern over humankinds energy requirements or the transition to so-called renewable energy sources. At present the problem lies within the framework of natural limits (*State of the World 1999*, 2000), whose assessment reveals that the power used by civilization has increased by an order of magnitude. Of course, the renewable sources of energy must be used, though an impartial analysis shows that all of them remain more or less ecologically dangerous. The main problem is to create the energy-saving technologies, to reduce an absolute value of consumed energy, and to protect natural ecosystems.

With respect to chemistry and the chemical industry, humankind behaves like the proverbial ostrich that sees no evil, hears no evil, and speaks no evil by burying its head in the sand. In December 2000, the 18-millionth substance was recorded, synthesized by man. The impact on human health of these substances has been studied for not more than 10,000–12,000 of them. Humankind has therefore transformed himself into an experimental rabbit, on which chemicals are tested by a "trial and error method" with unpredictable consequences. In Russian enterprises alone about a

billion potentially fatal doses of concentrated chemicals are produced each year
(*National Interests and Safety Problems of Russia*, 1997). Analysis of human tissue
in the USA has shown that it contains up to 2,000 pollutants (Colborn et al., 1996).
Especially dangerous are the so-called supertoxicants that damage the nervous and
endocrine systems, reduce reproduction, and cause breast and genital cancer.
Supertoxicants are non-threshold substances (i.e., they are dangerous in any concentration). As a rule, nature does not know these substances. They are chlorine-organic
pesticides, dioxin, biphenyl, plastics, etc. As a result, supertoxicants are accumulated
in the tissues of living organisms, including humans. Even minute doses in the environment can lead to high concentrations in the tissues of organisms at the end of
trophic chains. Estimates from Colborn suggest that concentrations can increase
millions of times for organisms at the top of the trophic chain. An estimate of the
damage to humankind from this chemical war against itself follows.

*These stereotypes are typical for technological optimists, but they are the result of
the main stereotype – the possibility of a reckless and unpunished destruction of natural
ecosystems.*

7.2 STEREOTYPE OF DESTRUCTION

The problem of the global ecological crisis should not be reduced only to the
problem of environmental pollution, which for many people remains a widespread
ecological stereotype. In fact, the main ecological problem that determines all the
others is natural ecosystem destruction due to economic activity (Kondratyev, 1990;
Gorshkov, 1995; Kondratyev et al., 2001b, 2003). Any economic activity, including
the tertiary economy, requires space and, hence, an unavoidable destruction of an
area of natural life. Trees and lawns in the cities and rural fields and pastures are not
natural ecosystems. All this is a part of an artificial environment created by man. Few
people ask the questions: what is the function of natural ecosystems in the system of
life, why do they exist for thousands of years, what does their destruction lead to?
Nevertheless, as far back as 1968, the outstanding Russian scientist Timofeev-
Resovsky answered these questions. He wrote that the biosphere of the Earth (i.e.,
the totality of all ecosystems) transforms energy and substances on the planetary
surface, forms the balanced composition of the atmosphere and of solutions in
natural waters, and through the atmosphere it is the life force behind the dynamics
of our planet. The same biotic mechanism affects the climate through the continental
water cycle of the globe, providing water evaporation by vegetative cover. Today,
with the advent of remote sensing by satellite, one may add the effect of vegetation on
the planetary albedo. Timofeev-Resovsky drew the conclusion that the Earth's
biosphere forms the human environment. Neglect and violation of its functions will
inflict not only damage on food resources and industrial raw materials but also on the
atmospheric gas and water environment (Tiuriukanov and Fedorov, 1996). As a
result, people will not be able to exist on Earth without the biosphere or with a
poorly functioning biosphere.

Thus the essence of the theory of biotic regulation and stabilization of the
environment has been clearly formulated. During the entire period of formation

and development of the Earth's civilization (see Chapters 4, 5 and 6), humankind has carelessly and pitilessly treated nature and all living organisms. This treatment was especially widespread in the 20th century (see Section 6) when a further 43% were added to the 20% of the land territory with natural ecosystems already destroyed during the earlier history of civilization. However, humankind treated itself in the same pitiless way. During the 20th century as many people were killed in wars as had been killed during the totality of all previous conflicts.

From the viewpoint of the economy, natural ecosystems can be considered a renewable resource and therefore they have a limited use (destruction) of their own. As Timofeev-Resovsky (1968) wrote, an injury to the gas and water environment and climate stability observed now is objective evidence. Timofeev-Resovsky (1968) and Schwartz (1976), being advocates of the ideas of modernism, proposed to increase the productivity of natural ecosystems artificially and to create stable cultural biogeocenoses. However, this proposition represents interference with the biosphere and an abuse of its proper function. During millions of years of evolution, the whole system – the biosphere or geographical environment – reached a dynamic equilibrium, but its so-called management has led this system towards an imbalance and ecological crisis of the planet. As for stable cultural biogeocenoses, including agricultural crops, they may be artificially created, but they will never self-produce without the participation of man, and the term "biogeocenosis" cannot be applied to them, since the task of biogeocenosis is not to meet the human needs but to maintain environmental stability. *If humankind wants to restore the damaged biogeocenosis to maintain environmental stability, then it will be better and cheaper to leave it to nature, which "knows" best and has powerful mechanisms useful for the reconstruction of natural ecosystems and biogeocenoses.*

Ecosystems are a set of sufficiently homogeneous correlated communities of organisms in balance with the environment. Such correlated communities are called biogeocenoses (see Chapter 3). They are similar to economic corporations in that they compete and oust those weaker communities that poorly regulate and stabilize the environment. Based on competition, the biosphere accomplishes both self-organization and management (see Chapter 3). The programme of competition is recorded in the genome of each organism, including humans, and the world's market economy is a realization of this basic law of life in our culture and civilization. But the results of this process in the biosphere and in the economy are quite different.

Similarities and differences between the economy and the biosphere (biota)

Similarities
Both the economy and the biosphere (biota) are based on competition (i.e., a free market). Concrete people (individuals) and corporative structures (bio-geocenoses) compete both in the economy and in the biosphere. The market economy and the biosphere are capable of self-organization: there is a strong relationship between supply and demand as well as prices in the

economy. In the biosphere an analogous balance exists, this time for nutrients – the substances needed for life.

Differences

Economy	Biosphere (biota)
Economic growth leads to an increase of physical mass, development for the sake of growth, and maximization.	Mass of biota remains constant as does the nutrients in the environment. Development is favoured over growth with a natural balance as the goal.
Expansion, capture, and broadening of space. Growth of corporative structures.	Expansion only for substitution of biogeocenoses not performing their functions, preserving the sizes of biogeocenoses and ecosystems.
Achievement of short-term goals only (maximum profit and maximum growth).	Long-term goals to preserve species (7 Myr), biogeocenoses, ecosystems and life on Earth.
Economy depends on nature and normal functioning of natural ecosystems.	Biota is self-sufficient, it is a true self-organized system optimizing the environment in its interests.
Economy is unstable.	Biota is stable. It is an example of sustainable development.
Economy ignores the boundaries and limits including those determined by biosphere laws.	Biota determines and observes the boundaries and limits: closed cycles of substances, distribution of energy by the size of consumers, ecological niches, and population density.

Biota use intraspecies competition to preserve, for as long as possible, the genetic programmes of species (i.e., to preserve the species itself). In other words, it is a means to maintain life stability. In the biosphere, the competition at the level of individuals of species and biogeocenoses in ecosystems plays the role of a mechanism to regulate and maintain the stability of the biosphere or geographic environment, and in the case of strong external forcing – reconstruction (change of the species composition and species structure of the biogeocenoses) and achieving a new level of regulation and stability. That is, the biosphere is a self-directed system regulating the environment on the basis of competition. Timofeev-Resovsky once said that there is no life on the Moon because there is no management there (Tiuriukanov and Fedorov, 1996).

The competition between individuals of the species *Homo sapiens* in existing socioeconomic conditions is directed not at preserving the genome but at raising prestige, thereby not ensuring species stability but favouring a disintegration of genetic programmes of humans and growing instability of the species. In the same way, the human management of the environment leads to its instability.

In the economy, the competition between corporations leads to expansion, growth of material flows, increased production, growth of prestige, and, as a result, growth of waste, destruction of ecosystems, and destabilization of the environment. By introducing a destabilization and chaos in the surrounding medium corporate structures exist and ensure their development. The main economic actors are now transnational corporations. Respectively, their role in this process of destruction of natural mechanisms of life security is growing.

A result of all this is, on the one hand, a disintegration of human genetic programs, humankind already achieving one-third of this lethal disintegration (Gorshkov and Makaryeva, 1997), and on the other hand, an accelerated destruction of ecosystems and elimination of organisms and species (a reduction of biodiversity). Thus, destruction, above a permissible threshold, of natural ecosystems formed during 4 Gyr of evolution destabilizes life on Earth at all levels – from molecular to global. *Therefore the main ecological problem for humankind is to preserve and reconstruct natural ecosystems in a volume which is sufficient to regulate and stabilize the environment, not the stereotypical struggle against pollution, itself an important problem but not the main problem on a global scale.*

A real model of such chaotic development of civilization in its relationship with nature on local scales had been realized in the past on Rapanui Island (Easter Island), where Polynesians populating the island finally destroyed the natural ecosystem, leading to an ecological and then socioeconomic catastrophe. The lands of our planet are no more than several islands, two-thirds of which have already had their natural ecosystems destroyed by man.

The destruction of natural ecosystems on Easter Island and its consequences

Easter Island (Rapanui) was discovered and populated by a small group of Polynesians in about the year 400 AD. At that time the island was covered with rich forests. The population was engaged in agriculture, fishing, and poultry breeding, gradually increasing in size. By the year 1500 the population reached 7,000. Trees were used to build dwellings, fishing boats, as fuel, and as rollers to move huge stone idols (the number of which exceeded 1,000). Therefore, by this time, the forest resources had been practically destroyed, resulting in soil erosion and desiccation. As a result, agricultural production diminished. There was a deficit of wood to build boats and dwellings. The ecological crisis led to a food crisis, and then a sociopolitical crisis: humans moved to caves for shelter, fishing stopped, waring for resources among groups occurred, and even cannibalism appeared. Jacob Roggeveen, who discovered the island in 1722, saw only 500 people

> dragging out a miserable existence. They did not remember their history and
> did not know why so many stone idols were on the island (Tickell, 1993).

The threshold of ecosystem destruction can be determined by the energy distribution of the number of organisms in natural ecosystems (see Figure 3.1) that have
been obtained from a generalization of the vast amount of empirical material
(Gorshkov, 1995). Vernadsky (1944) wrote that a human cannot independently
build his history regardless of the laws of the biosphere. The global ecological
problems appearing in the 20th century are the result of such a free building of
history. Such a method is now assumed to be obsolete. *It is necessary to build a
new history of civilization, which would agree with the laws of the biosphere. By the
highest standards, it will be a continually maintained or, as is usually written in our
scientific and popular literature, sustainable development. In fact, one should speak not
about sustainable development but about sustainable living – continually maintained
life, as was suggested in the book* (*Caring for the Earth*, 1991). It is the last term that
corresponds to the goal declared in the report of the Brundtland Commission,
namely, *to satisfy the needs of living people without depriving next generations of
the possibility to satisfy needs of their own* (*Our Common Future*, 1989). It means,
as a minimum, future developing the world community, providing it with a better
and higher level of food and comfort, leading to the notion of "sustainable development". Clearly, the present means of civilization development does not help to
achieve this goal. In spite of impressive growth in the economy in the 20th century
(see Section 6.5), the present socioeconomic system has not resolved the issues of
hunger, poverty, access to potable drinking water, infectious diseases, etc. At the
same time, this economic growth has destroyed a basic part of ecosystems and
emitted into the environment large amounts of pollutants, far exceeding ecological
limits.

7.3 MYTHS OF THE MODERN CIVILIZATION

The stereotypes mentioned above are the result of modernistic ideology formed
during the last 300 years. Based on assimilation and analysis of information,
people generate new information (including ideologies and doctrines) passing it
over to future generations. Along with correct information they often create
incorrect information. The history of civilization is mainly a realization of
incorrect information (or at least information initially believed to be correct but
later proven incorrect). In the 20th century, in light of the civilization–nature confrontation, many stereotypes generated by modernism were incorrect (or later
proved incorrect). Apart from stereotypes, modernism had earlier originated and
secured myths, which originally had been incorrect. They may be called myths of
civilization.
    Modernism had put a human being at the centre of the world and announced
themselves the goal and the highest achievement of evolution. In principle, humans

had done this at the dawn of civilization by humanizing their gods. Anthropo-centrism has no scientific basis. Genetics has proved that evolution is a stochastic process, and humankind is only one of a large quantity of species of organisms. This myth had been useful to develop civilization before its confrontation with nature, but now it should be rejected. It is life materialized in biota (with human beings as part of it) that should be placed at the centre of the world.

Another myth (connected with the previous one) is that humans can use biospheric resources with impunity and at their whim. This myth had been reflected in the Bible and then supported by science at the dawn of modernization when Francis Bacon (1561–1626) announced "knowledge is power". In the 19th and 20th centuries the main motive following from this slogan became "conquering the nature". But in the second-half of the 20th century there were many who understood that knowledge is not simply power but a dangerous power with unpredictable consequences. Now it became clear that human beings use knowledge to more actively consume resources and energy – to destroy nature and extract more and more resources. In places of destroyed natural communities an artificial environment is created with a single goal – to provide comfort for humans.

One further myth, still popular, is that of inexhaustible natural resources. As far back as the early 20th century, Malthus wrote about a limit to or threshold of development, and in the 20th century the Club of Rome (the community of world sceintists) revived this idea. However, the resource threshold can be changed, for instance, by using another resource for some time based on knowledge and tech-nologies (but not indefinitely). The threshold, which has already been reached, is the ecological limit, which can be neither moved nor substituted with the help of knowledge and technology.

Solution of ecological problems by technological means is one further myth in the spirit of modernism. The experience of solving a seemingly simple problem of pollution has shown that enormous expenditures in billions of dollars on protection and cleaning of the environment, new technologies, and resource saving have made it possible to carry out only local cleaning of the environment. In this case the ecolo-gical balance of these local clean-ups turned out to be negative: cleaning one local site leads to pollution of another site according to the law of conservation. One can say that "sweeping litter under the carpet" takes place, but often under someone elses' carpet, when dirty and resource-consuming products are moved from developed to developing countries. (Danilov-Danilyan and Losev, 2000). *Develop-ment of a technical system able to regulate and stabilize the environmental character-istics quasioptimal for life on global scales with an accuracy of biogeocenoses, ecosystems, and the biosphere is unlikely* (Figure 7.1) (Gorshkov, 1995; Gorshkov et al., 1996, 2000).

Thus the stereotypes existing in the economy and ecology and myths in human minds as well as ethical values formed on this basis contradict the requirements of life stability. If a primitive human could answer the question about the meaning of life they may say it means the stability of the family, while a more civilized human may specify the stability of the state. It is now clear that one should add a provision of the stability of life and environment on the planet, without which neither the

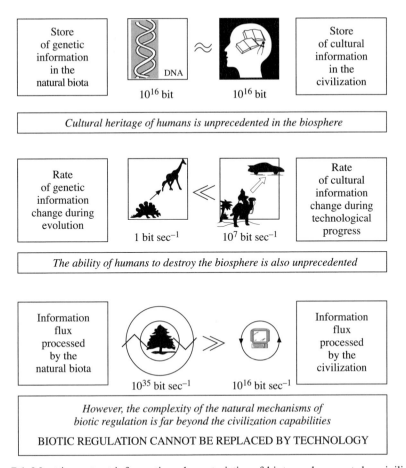

Figure 7.1 Most important information characteristics of biota and present-day civilization.

human race nor the states stability can be preserved. Provision of life and civilization stability and their sustainable development are only possible if they are governed on the basis of the laws of the biosphere (i.e., on a scientific basis). The present market economy cannot realize this, since it has another goal – to obtain a maximum profit, which is more rapidly achieved in conditions of intensified chaos and instability. However, we do not call to eliminate the market economy. It is necessary to ensure the innovation process. The governing decisions should be aimed at determining the ecological and moral "boundaries" for the market, within which it should not be limited.

    All the enumerated stereotypes and myths are also characteristic of Russia (Kondratyev et al., 2001b). They are very persistent for three reasons:

1   during the 70 years of the existence of a centralized system of government the domination of nature was paramount;

2   the vast territory of the country implies huge resources; and
3   the standard of living in developed countries is a strong incentive for people to
     follow the same manner of destruction of natural ecosystems.

Russia is a large country, greater than Australia, Antarctica, and Europe, but it is located in the cold north-eastern "corner" of Eurasia. About 8 million people live in the severe region called the North, which covers two-thirds of the Russian territory. However, this region has preserved vast tracts of natural, mainly forest, ecosystems. It is the largest tract of virgin and partially virgin nature in the world. It is this that constitutes the world value of the northern territory of Russia. It is impossible to estimate its value, since this is the foundation of life and life itself. But some of its functions can be evaluated. One of these functions is a removal of anthropogenic carbon dioxide from the atmosphere and its deposition.

According to the Kyoto Protocol, the developed countries started reducing carbon emissions to the atmosphere down to a certain level. The minimum cost of emission reduction per 1 t of carbon constitutes from US$550 to 1,100 (Bedritsky, 2000). Natural ecosystems preserved in Russia, according to our estimates, apart from complete removal of emissions of the country itself, absorb as a minimum, an additional $300 \, \text{mt} \, C \, \text{yr}^{-1}$ (i.e., every year they remove from the atmosphere and deposit, for the most part, carbon emitted by the developed countries with a financial worth of US$160–325 billion). The lower value of which represents Russia's foreign debt. One could argue that Russia invests this sum in Western countries. During a 10-yr period this sum would constitute US$1.6 trillion. From available estimates (Putin, 2000), the price of the profitable part of all mineral resources of Russia now constitutes US$1.5 trillion (for details see Chapter 8). Important economic and ecological conclusions follow from these data and estimates.

The strategy of developing territories and allocation of production in the country should be changed radically in Russia. Development of vast cold and unfavourable regions with the practical absence of infrastructure and large expenditures on energy and transport is economically unprofitable, since the value of natural ecosystems will increase with their global reduction. The economy of Russia should be more compact, concentrated in the climatically comfortable regions of the middle and southern band of the European part, southern Siberia and Far East, where the main part of the population lives, with a sufficiently developed infrastructure and labour force. The rest of the territory should be left to numerous native peoples traditionally living in these regions. Some minerals should be extracted, with mining remaining if profitable or strategically necessary.

The development of projects of nature transformation should be stopped as well as of great constructions like the north-to-south transfer of river flows, driving a tunnel between the mainland and Sakhalin Island, building a railway to the Bering Strait, etc.

The slogan "develop and conquer" should be rejected, bringing forth the task to raise the efficiency of the economy and to preserve natural ecosystems, which correspond to the requirements of sustainable life and development within the limits permitted by biospheric laws. Businessmen should solve the first problem with governments solving the second.

The time since the United Nations Conference on Environment and Development (UNCED) in Rio de Janeiro has shown that the spirit of UNCED has gradually disappeared, having been ousted by the globalization and liberalization of the market. Nevertheless, in our opinion, the main cause behind the blurring of the idea of sustainable development is because the large expenditures on environmental protection, serious global nature protection infrastructures, and nature protection legislation developed in many countries is not effective. It was stated at the Rio + 5 Conference that global ecological stress is growing. The same statement was made at the Rio + 10 Summit (Kondratyev et al., 2002b).

The main reason for such a situation is that "Agenda 21" contains nothing more than its usual measures. National programmes and strategies of sustainable development have been developed by the same principle (Danilov-Danilyan and Losev, 2000). But in conditions of the civilization–nature confrontation and the war humankind is waging against the rest of life, there is no chance for the transition to sustainable life with usual measures, nor for prevention of an ecological catastrophe. In fact, "Agenda 21" has no scientific basis. It is based on observations viewed in a very narrow and one-sided manner and on the earlier experience of humankind, which is then projected onto the future. But in new conditions of the civilization–nature confrontation the earlier facts and experience should be interpreted from the scientific point of view. Only the scientific basis of sustainable life – the theory of biotic regulation of the environment – can provide this interpretation, and not only with respect to solution of ecological problems.

8

The carbon budget, role of biota, and ecosystem cost

8.1 SOURCES OF ANTHROPOGENIC CARBON

The world community has long been concerned about the process of warming the near-surface layer of the atmosphere due to the "greenhouse effect" – generated by anthropogenic emissions of carbon dioxide and other greenhouse gases. However, a complete account of anthropogenic emissions including carbon dioxide has not yet been made. Special attention has been paid to industrial emissions resulting from fossil fuel burning, with carbon emissions due to land use being neglected.

First of all, it should be mentioned that over long time periods of tens of millions of years, fluctuations in the carbon dioxide concentration have been held within narrow limits – an order of magnitude. During the last million years these fluctuations remained in a narrow range.

Variations of carbon dioxide between the interglacial and glacial epochs

Studies of ratios of boron isotopes in the shells of ancient foraminifers (Pearson and Palmer, 2000) have made it possible to assess the acidity (pH) of the World Ocean surface layer, and through this to reconstruct the CO_2 content in the atmosphere during the last 60 Myr. They showed that during this period, changes of its concentration did not exceed one order of magnitude. In the late Palaeocene and early Eocene (60–52 Myr ago) (i.e., on the ice-free warm Earth), CO_2 concentration in the atmosphere was within 2,000–4,000 ppmv CO_2). In the early Miocene (24 Myr ago) with the beginning of glaciation it decreased to 500 ppmv.

Judging from the results of analysis of an ice core from the Antarctic station "Vostok" (Danilov-Danilyian and Losev, 2000), during the last 420,000 years the CO_2 level on the Earth survived cycles of oscillations coinciding with the stages of advance and retreat of glaciers. The CO_2 content changed within 180–200 ppmv during the glacial periods to 265–280 ppmv during interglacial periods. Palaeodata show a close correlation between air temperature and atmospheric CO_2. This suggests the conclusion that the latter is a primary driver of climate change at the end of the period of glacial propagation. Cyclic periodicity strictly corresponds to the cyclic oscillations of the Earth's orbital parameters (the Milanković cycles). However, it was proved that the accompanying changes of insolation are insufficient for large-scale climate changes, however, they may be a triggering mechanism for a system of the control of CO_2 content, which is determined by interaction between land and ocean biota.

The present carbon dioxide content has increased during the 20th century from 280 to 360 ppmv, which is almost equal to the range of CO_2 oscillations between a maximum of the last glaciation (190 ppmv CO_2 18,000 years ago) and an optimum of the Holocene (280 ppmv CO_2) (i.e., the rate of the present CO_2 change is 100 times larger).

A rapid growth of coal extraction and then of oil and gas began in 1850. From this time on, the atmosphere continually received CO_2 emissions from fossil fuel burning (the CO_2 mass is recalculated into carbon by dividing it into the coefficient 3.664). For the last 100 years the carbon content in the atmosphere increased by 30%. An especially sharp increase due to industrial emissions (a factor of 4.6) was registered between 1950 and 1996. From 1996 (in this year, 6.52 Gt C were emitted into the atmosphere (*Global Environment Outlook 2000*, 1999)) global emissions continued to grow. On the whole, during the last 20 years emissions have increased by 38%. Their reduction during this period occurred twice: in 1981–1982, probably, in connection with the use of energy-saving technologies in the developed countries, and in 1992–1994 both as a result of reduced energy consumption in the countries which began to reject the old economic systems, and as a consequence of the growing use of natural gas instead of coal and oil in European countries and in Russia. Unfortunately, these processes had only a temporary effect on the dynamics of emissions. The general trend remained unaffected.

From 1980 until 1990 the annual increase of carbon emissions averaged 84 Mt. Between 1991 and 1996 this increment was somewhat lower – 83 Mt. Thus the global emission in 2000 reached 6.85 Gt (*Global Environment Outlook 2000*, 1999). If fossil fuel burning continues at the same rate, the content of CO_2 will double by 2060, which, according to the Intergovernmental Panel on Climate Change (IPCC) may cause an increase in the average global temperature of between 1.4°C to 4.8°C.

In Russia, after the USSR disintegration, the same trend has remained – hushing up information about emissions of greenhouse gases. Reports on the state of the environment lack information about industrial carbon emissions. From indirect data and taking into account international estimates, Losev and Ananicheva (2000) calculated emissions of carbon into the atmosphere as of 1991 constituted 600 Mt. In 1994, due to a decrease in production, emissions were reduced to 480 Mt. From

Table 8.1. Emissions of carbon in Russia due to fossil fuel burning in 1994 taken from different sources.

Source	Emission of carbon ($Mt\,yr^{-1}$)
State of the World 1996 (1996)	455
UES (United Electrical Systems) of Russia and Institute of Sustainable Communities	500–550
IPCC	437
Klimenko et al. (1997)	532*
Losev and Ananicheva (2000)	480

the book *State of the World 1996* (1996), in 1994 emissions of carbon in Russia constituted 455 Mt. Table 8.1 contains estimates of CO_2 emissions in 1994, from different sources.

From Table 8.1, the industrial emission of carbon in Russia in 1994 averaged 475–485 Mt, but the basic figure for 1994 can be considered 455 Mt (*State of the World 1996*, 1996), with this figure being the basis for further estimates.

One means to assess the industrial emission of carbon is to calculate it from the dynamics of emissions of sulphur dioxide and nitrogen oxide, since burning of natural fuel is followed by an emission of these substances. Losev and Ananicheva (2000) used this method to estimate emissions of carbon in 2000. According to the book *On the Environmental Condition* ... (1998), in 1994–1998 emissions of sulphur dioxide decreased by 8%, nitrogen oxide by 9.5%, and solid particles by 21%. With an assumed minimum value of carbon emission (8%) and 455 Mt as an initial estimate of emissions in 1994 (*State of the World 1996*, 1996), its absolute value in 1998 constituted 420 Mt C (representing a per capita value of $2.9\,t\,C\,yr^{-1}$ in Russia). Emissions of carbon in 2000 should be approximately the same as in 1998, since in 1997 and 1998 there was a trend of stabilization of emissions of accompanying substances, with a decrease of production in 1998 compensated for by the growth in the second-half of 1999 and in 2000.

For a better-founded estimate of industrial emissions of carbon in Russia, the data on per capita energy consumption in 1999 was used (Shelepov, 2000), which was 6 t c.f. (conventional fuel). Bearing in mind that 10% of the energy was provided by hydroelectric and atomic power stations, the per capita fossil fuel consumption constitutes 5.4 t c.f. The resulting structure of fossil fuel consumption in the country is as follows: 19% – coal, 23% – oil, 58% – gas. The per capita emission of carbon in Russia was calculated based on these data, using the modified Marland-Rotti method (Klimenko et al., 1997). It follows that the total industrial emission of carbon in 1999 was 400 Mt. This estimate is very close to 420 Mt in 2000 (Losev and Ananicheva, 2000). Taking into account some growth of the economy in 2000, one can state that carbon emissions by the beginning of the 21st century in Russia reached $400–420\,Mt\,yr^{-1}$.

With growing industrial activity, CO_2 emissions will inevitably increase. So far, being at a transition period, Russia suffers intensive emissions from industrially developed and rapidly developing countries that border it both in the east – China,

Table 8.2. Emissions of carbon from 20 countries in 1994 (*State of the World 1996*, 1996)

Country	Total emission (Mt)	Per capita emission (t)	Emission per dollar of GDP (t)	Growth of emmissions in 1990–1994 (%)
USA	1,371	5,26	210	4.4
China	835	0.71	330	13.0
Russia	455	3.08	590	−24.1
Japan	299	2.39	110	0.1
Germany	234	2.89	140	−9.9
India	222	0.24	160	23.5
England	153	2.62	150	−0.3
Ukraine	125	2.43	600	−43.5
Canada	116	3.97	200	5.3
Italy	104	1.81	110	0.8
France	90	1.56	80	−3.2
Poland	89	2.31	460	−4.5
South Korea	88	1.98	200	43.7
Mexico	88	0.96	140	7.1
South Africa	85	2.07	680	9.1
Kazakhstan	81	4.71	1,250	–
Australia	75	4.19	230	4.2
North Korea	67	2.90	960	–
Iran	62	1.09	270	–
Brazil	60	0.39	70	15.8

* GDP data are taken for 1993 and obtained with the purchasing capacity of the population taken into account.

Japan, the Koreas – and in the west – European countries including the former republics of the USSR (Table 8.2).

The total emissions of carbon in European countries in 1994 (1,400 Mt) is equivalent to the emissions of the greatest emitter in the world – the USA. In 2000 the USA emitted more then 1,600 Mt C (Figure. 8.1). According to EPA (Environmental Protection Agency) data, the annual increment for the next 10 years will constitute about 20 Mt, after which it will decrease a little. However, in connection with the programme of energy development after the energy crisis in California, a more intensive growth of CO_2 emissions in the USA is possible.

As for the contribution of Europe, at the turn of the century the quantity of carbon emissions decreased: in the countries with transition economies it either decreased or stabilized (due to economic depression). The decrease from developed countries was as a result of goal-directed energy policy. The comparative contribution of Russia to total emissions of carbon from fossil fuel burning in 2000 can be judged from the data of Table 8.3.

By the beginning of 2001 the developed European countries, former socialistic countries, and the USA are responsible for almost half the emissions of industrial carbon. Added to China, Japan, and both Koreas it increases to 65%, finally with the contribution of Russia it increases to 70% of the total world emissions.

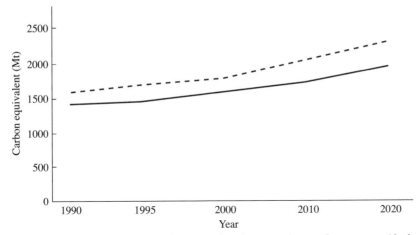

Figure 8.1 The current emission and forecast of anthropogenic greenhouse gases (dashed line) and carbon (solid line) for 2020 in the USA.

Source Environmental Protection Agency.

Table 8.3. The comparative contribution of Russia to the global industrial emission of carbon in 2000.

Regions and countries	Industrial emission (Mt)	Percent of global emission
Europe (without Russia)	1,400	20
China, Japan, and countries of Korean Peninsula	1,400	20
USA	1,625	24
Russia	420	6
The World	6,850	100

There is another, not less important emission of carbon emitted due to the destruction of ecosystems as a result of economic (mainly agrarian) activity and deforestation. Estimates of this at the beginning of the 1990s are given in Table 8.4 (*Global Environment Outlook 2000*, 1999).

The existing estimates of the global emission due to land use obtained by different authors constitute: $2.4\,Gt\,yr^{-1}$ (Tarko, 1994), $1.4\,Gt\,yr^{-1}$ (Klimenko et al., 1997), $1.6\,Gt\,yr^{-1}$ (Houghton et al., 1996), and $2.4\,Gt\,yr^{-1}$ (*Protecting the Tropical Forests*, 1990). They differ markedly from each other and substantially from the data in Table 8.4. It is still unclear, whether these figures are net or total emissions of carbon. We believe that an inclusion into Table 8.4 of agricultural waste is not justified, since the CO_2 flux from this source is partially compensated for when a new yield is ripening in the fields. With this component excluded, the total emission from biota destroyed due to economic activity will constitute $3,030\,Mt\,yr^{-1}$, which is much closer to the estimates given above. However, for all estimates of emissions due to land use the error of their estimation remains unknown, including the data of

Table 8.4. Global emission of carbon due to biomass destruction by the beginning of the 1990s.

Type of biomass	Destroyed biomass (Mt of dry substance)	Carbon emission ($Mt\,yr^{-1}$)
Savannah	3,690	1,660
Agricultural waste	2,020	910
Tropical forests	1,260	570
Wood burning	1,430	640
Temperate boreal forests	280	130
Charcoal	20	30
Total	*8,700*	*3,940*

Table 8.4. Summing the global values of carbon emissions due to fossil fuel burning with the emissions due to deforestation and land use, we obtain the total emission for the year 2000 as being equal to $9{,}880\,Mt\,yr^{-1}$. This value shows that it should not be neglected, as is often done in different calculations, including the use of climate models.

In Russia, deforestation and forest fires contribute most to this kind of emission. The areal extent of the economically developed forests of Russia is about 250 Mha. Before the 1990s, up to $420\,million\,m^3$ of forests were cut every year in Russia. With the remains from clearings (roots, branches, etc.) taken into account, the total amount of carbon as phytomass of the felled trees was estimated at $165\text{--}170\,Mt\,C\,yr^{-1}$ (Isaev and Korovin, 1999). From the estimates by these authors, by the year 1989 the total flux of carbon (destruction of transported wood and waste burning) in connection with cuttings constituted $450\,Mt\,CO_2$ or $123\,Mt\,C\,yr^{-1}$. In 1993 the emission decreased to 73.6 Mt, and in 1995–1997 to $60\text{--}65\,Mt\,C\,yr^{-1}$, as industrial cuttings decreased.

Forest felling both for export and for new construction has recently intensified. Judging from official statistics, $98\,million\,m^3$ of wood were cut in 1998. Ten years ago the need for firewood in Russia constituted 20% of felling, which reached $336\,million\,m^3$ (Losev et al., 1993). During the last decade this requirement decreased (i.e., in 2000 it constituted no less than $60\text{--}70\,million\,m^3$).

With illegal felling taken into account, the authors believe that up to $200\,million\,m^3$ of timber are now cut in Russia. A substantial part of this is transformed into carbon emissions (wood burning, and short-lived products), which constitutes $40\,Mt\,C\,yr^{-1}$. This is an upper limit of the estimates, since a considerable portion of the wood is exported and some is deposited in the form of long-lived products (furniture, building materials, etc.). Emissions due to wildfires should also be added.

In the Asiatic part of Russia, forest fires contribute significantly to carbon fluxes. Only two-thirds of forests are actively protected. In these territories a monitoring of the consequences of wildfires is carried out as well as an accounting of the areas of burning. Every year, 10–35,000 forest fires are recorded over the 0.5–2.2 Mha

Table 8.5. Sources of carbon emission due to forest fires in Russia
(Mt C yr).

Sources	Emission
CO_2 and soot from burning vegetable materials	15–45
Destruction of dead but not burnt vegetation	20–53
Total emission	35–93

territory (Isaev and Korovin, 1999). In general, different authors estimated the total
area of fires in boreal forests at 2.0–12 Mha yr^{-1} (Konrad and Ivanova, 1998; Isaev
and Korovin, 1999). As a result of forest burning, various sources emit carbon into
the atmosphere (Table 8.5).

Emission of carbon due to soil sources as a result of land use and soil erosion can
be estimated (Tarco, 1994), on a global scale at 1 Gt C yr^{-1}. For the global area of
agricultural land of 30 million km^2, we calculate that 33 Mt C fall on 1 million km^2
of land. Since the area of agricultural land in Russia is 2.2 million km^2, emissions of
carbon from this land can be estimated at 70 Mt yr^{-1}.

Thus in the year 2000, emissions of carbon due to fossil fuel burning in Russia
constituted 420 Mt C, forest felling 40 Mt C, forest fires 35–93 Mt C, soil erosion
70 Mt, and total anthropogenic emission 565–623 Mt C yr^{-1}. Thus the share of
Russia, according to obtained estimates, in the global total emission of carbon
due to all kinds of economic activity, constitutes 6%.

There is another method for estimating the emission of carbon due to ecosystem
destruction – based on the law of conservation and stoichiometric ratios of oxygen
and CO_2 observational data (Gorshkov and Makaryeva, 1998; Gorshkov et al.,
2000). Calculations using this method have shown that emissions of carbon by
land biota from territories where, through economic activity humankind has
broken the natural stability, constituted 6.3 Gt yr^{-1} during the period 1991–1994.
This figure is the total emission of carbon by disturbed global biota. One hectare of
developed land emits 78 kg C yr^{-1} into the atmosphere. This is 40% greater than the
estimate given in Table 8.4 and higher than emissions due to fossil fuel use. With the
estimate 6.3 Gt yr^{-1} assumed, the total global anthropogenic emission of carbon in
2000 constituted about 13,150 Mt yr^{-1}.

*Thus there are two powerful anthropogenic sources of carbon: destruction of
natural ecosystems and soils as a result of agrarian and industrial activity, and fossil
fuel burning. The first source, is more significant than the second one, but it is the
second one upon which the attention of the scientific community is mainly concentrated.*

8.2 BUDGET OF ANTHROPOGENIC CARBON

Estimates of the mass of anthropogenic carbon emitted from its sources into the
environment differ over their accuracy. If carbon emissions due to fossil fuel burning

are accurately estimated, and the error is below 10%, then estimates of pure emission due to land use are rather uncertain. It should be calculated as a difference between total emission due to economic activity and assimilation by preserved, partially destroyed, and undestroyed ecosystems, which serve as sinks for excess carbon. Houghton (1997) noted that there are five methods to estimate these quantities:

1   direct measurement of the change of carbon amounts in land ecosystems;
2   direct measurements of the CO_2 fluxes over the surface of the ecosystems;
3   modelling the changes of productivity and respiration due to changes of the environment;
4   modelling the global carbon cycle from geochemical data; and
5   modelling the changes of land carbon as a result of land use.

The first 3 methods are complicated, expensive, and incur significant errors. The last 2 methods, in Houghton's opinion, enable estimation of carbon fluxes on global scales.

    An assessment based on the carbon cycle and the geochemical data shows that the land ecosystems are a pure *sink* of carbon, ranging between 0.3 and 1.86 Gt yr^{-1}. Analysis of land use, in turn, shows that it serves as a pure *source* of carbon equal to 1.6 ± 0.7 Gt yr^{-1}. If both analyses are valid, then the total mass of carbon assimilated by non-destroyed natural and reviving ecosystems is equal to the sum of these estimates (i.e., from 1.9 to 3.4 Gt yr^{-1}), with the tropical forests being the main source of emissions, and the temperate forests a sink (Houghton, 1997). So, the difference is significant. At the same time, a method has been proposed to calculate the rate of carbon change within its reservoirs based on the law of conservation (Gorshkov, 1995; Makarieva, 2000).

    There are five basic global reservoirs in which carbon is redistributed due to human activity: fossil fuel, dissolved inorganic carbon in the ocean; dissolved organic carbon in the ocean; organic carbon of land biota, and atmospheric carbon. We will denote them with indices 1, 2, 3, 4, and 5 respectively. For the period 1970–1990, from analysis of changes of the ratio of carbon isotopes $^{13}C/^{12}C$ in the surface waters of the World Ocean, the rate of carbon assimilation from the atmosphere by the physico–chemical system of the ocean was estimated at 2.2 ± 0.4 Gt yr^{-1} (Keeling et al., 1996; Gorshkov et al., 1998; Makarieva, 2000).

    During the period 1991–1994, direct measurements were made of the rate of change (extraction and burning) of carbon supplies in fossil fuels, giving 5.9 ± 0.15 Gt yr^{-1}. Since in the preindustrial period the concentration of carbon in the atmosphere remained practically constant, the rate of its assimilation from the atmosphere by the physico–chemical system of the ocean should increase in proportion to an increment of carbon content in the atmosphere with respect to an equilibrium preindustrial value. The mass of carbon in the atmosphere has changed from 130 (averaged over 1970–1990) to 170 Gt yr^{-1} (averaged over 1991–1994). Hence, the average rate of assimilation of inorganic carbon by the physico–chemical system of the ocean constituted 2.2 ± 0.4 Gt yr^{-1} for this period. Multiplying this figure by (170 : 130), we obtain 2.6 ± 0.5 Gt yr^{-1} – an average value of assimilation by the physico–chemical system of the ocean between 1991 and 1994. From the data on the growth of CO_2 concentration in atmosphere, 2.2 ± 0.1 Gt C accumulated every year during the period 1991–1994 (Gorshkov, 1995; Gorshkov et al., 2000).

Thus the rate of change of the carbon supply (m) in three reservoirs (atmosphere, fossil fuel, and inorganic carbon of the ocean assimilated from the atmosphere) is known with high accuracy. The rate of change of the mass of carbon in the remaining two reservoirs (land biota and dissolved inorganic carbon in the ocean) can be found from the law of conservation, since the sum of all rates of carbon change in these reservoirs is equal to zero, that is:

$$m_1 + m_2 + m_3 + m_4 + m_5 = 0, \tag{8.1}$$

with all sources of carbon in the equation having a positive sign and all sinks, a negative sign.

Both synthesis and disintegration of organic carbon are followed by opposite changes of the mass of oxygen. A change of the oxygen content in the atmosphere has been measured, and in the rest of the reservoirs it can be connected with a change of carbon content in them through the known stoichiometric ratios of O_2/CO_2 (a_1, a_2, a_3, a_4, a_5). Hence, we obtain Eq. (8.2), the oxygen balance in four reservoirs:

$$a_1 m_1 + a_2 m_2 + a_3 m_3 + a_4 m_4 + a_5 m_5 = 0. \tag{8.2}$$

Stoichiometric ratios are obtained from empirical data and for the oceanic pool a_3 constitutes 1.30 ± 0.03 (the Redfield ratio), for the fossil fuel pool $a_1 = 1.38 \pm 0.4$, for the pool of land biota $a_4 = 1.10 \pm 0.05$, and for the atmospheric pool a_5 (from measurements by Keeling et al., 1996) constitutes 2.2 ± 0.2 (Gorshkov et al., 2001). Since the stoichiometric ratio with carbon assimilation by the physico–chemical system of the ocean is not followed by changes of the atmospheric oxygen concentration, it ($a_2 m_2$) is equal to zero. From these two equations, for the period 1991–1994 the rate of change of dissolved organic carbon in the ocean is 4.9 ± 0.7 Gt yr^{-1}, the pure emission of carbon due to land use is 3.8 ± 0.6 Gt yr^{-1}, and the total emission from land ecosystems is 6.35 Gt yr^{-1}. Thus the rates of change of the carbon supply in all active reservoirs, total emission of carbon by land biota due to land use (6.3 ± 1.0 Gt yr^{-1}), and carbon sink for preserved natural ecosystems on land (2.5 ± 0.4 Gt yr^{-1}) were obtained with a sufficient accuracy (for details see Gorshkov et al., 2000).

Other reservoirs of carbon are inert, and their source and sink values lie within the errors of calculations for the basic reservoirs. Gorshkov et al. (2000) believe that the error in determination of the terms of Eqs (8.1–8.2) does not exceed 10%. The basic global sources, sinks, and respective balances are given in Table 8.6.

Now, one can use these data to estimate the carbon balance over the territory of Russia. In Section 8.1 the emission of carbon due to fossil fuel for 2000 is estimated at 420 Mt yr^{-1} in Russia. It decreased compared to 1991–1994 values, whereas the global emission increased to 6,850 Mt yr^{-1}. Nevertheless, the data in Table 8.6 were taken as the basis for calculations. This is not a serious assumption since the calculated quantities in the second-half of the 1990s were conservative and changed negligibly by amounts not exceeding the errors of the calculations.

Table 8.6. Global sources and sinks of carbon for the period 1991–1994 $(Gt\,yr^{-1})$.

Sources	Sinks
Fossil fuel burning (5.9)	Land ecosystems (2.5)
Land use (6.3)	Physico–chemical system of the ocean (2.6)
	Atmosphere (2.2)
	"Biological pump of the ocean" (4.9)
Total = 12.2	*Total* = 12.2

The second assumption, also insignificant, is that the carbon emission due to land use per unit of territory of Russia is assumed to be close to average global values owing to the fact that the productivity of the ecosystems of Russia is close to the average global value.

The share of global land area on which natural ecosystems are either destroyed or strongly deformed constitutes 63%. This represents 85 million km^2, if the global land area is taken to be 135 million km^2 (excluding areas covered with glaciers and rock). The area of such territories in Russia is 6 million km^2, or 4.5% of the global area with an active land use, emitting $6,300\,Mt\,C\,yr^{-1}$. Hence, the share of Russia is $6,300\,Mt\,yr \times 0.045 = 285\,Mt\,yr^{-1}$. Thus Russia is a source of anthropogenic carbon equal to $420 + 285 = 705\,Mt\,yr^{-1}$, where 420 Mt are total industrial emissions due to fossil fuel burning, and 285 Mt are total emissions due to land use (i.e., greater than the value calculated from expert estimates, see Section 8.1).

The physico–chemical system and the biological pump of the World Ocean (dissolved organic carbon) together assimilate $7,500\,Mt\,C\,yr^{-1}$. In Russia, the share of carbon assimilated by the ocean should be proportional to the share of the total emission of the country with respect to the global value $(12,200\,Mt\,yr^{-1})$ (i.e., 6%). Hence, the value of the Russion emission of carbon absorbed by the World Ocean constitutes $7,500\,Mt\,yr^{-1} \times 0.06 = 450\,Mt\,yr^{-1}$ (Table 8.6).

Russia has an enormous mass of undisturbed, weekly disturbed, and reviving natural ecosystems, constituting 22% of the global natural mass (50 million km^2) which are the main areas of CO_2 sinks (Danilov-Danilyan and Losev, 2000). The sink of carbon into this reservoir can be calculated as follows: $2,500\,Mt\,yr^{-1} \times 0.22 = 550\,Mt\,yr^{-1}$ (Table 8.6).

As a result, in Russia the total sink of carbon into the World Ocean and its remaining (mainly forest) ecosystems constitutes $1,000\,Mt\,yr^{-1}$, which is greater than its total emission by $295\,Mt\,yr^{-1}$. Hence, Russia is an area acting as a pure sink of anthropogenic carbon. *Russia introduces nothing into the growth of CO_2 concentration in the atmosphere, and, on the contrary, restrains this growth due to its preserved natural ecosystems.*

The calculation given above supposes a certain error underestimating the values of carbon sinks for reservoirs in the Russian territory because of an assumption that forests (which are the main regulators of carbon fluxes) in the world have similar potentials of assimilation of anthropogenic carbon. At the same time, virgin boreal forests have the highest potential, as well as completely preserved natural ecosystems. They function as mechanisms of stabilization and regulation of the environment (Gorshkov et al., 2000; Houghton, 1997). The scientists of the Institute of World Resources (Bryant et al., 1997) have made an inventory of such forests. According to their expert data, the share of virgin forests in Russia constitutes 26% of the area of all forests of this type. The quantitative data on the degree of violation of the ecosystems in different biogeographic zones and the data of the forest inventory in Russia for 1998, with their age taken into account, enable us to estimate the area of virgin and quasivirgin forests (aged 100 years and more and retaining the closure of the cycles of nutrients) as half the area of all forests of Russia or one-third of the global forests (Hannah et al., 1994; Losev et al., 2001a). From these estimates of the share of the area of virgin forests of Russia one can obtain two estimates of the carbon sinks: $2,500 \, \text{Mt} \, \text{yr}^{-1} \times 0.26 = 650 \, \text{Mt} \, \text{yr}^{-1}$ and $2,500 \, \text{Mt} \, \text{yr}^{-1} \times 0.33 = 825 \, \text{Mt} \, \text{yr}^{-1}$. Respectively, the total sink (with the sink into the ocean taken into account) should be, 1,100 and $1,275 \, \text{Mt} \, \text{yr}^{-1}$.

Figure 8.2 is a graph of the state of Russian forests, from which it can be seen that with the natural reconstruction of forests, cut by humans and burnt in fires, they have an enormous potential to assimilate carbon – the upper dashed curve corresponding to the state of persistent natural ("wild") forests. At the same time, plantation forest (silviculture) and agrosystems never reach such capabilities to assimilate carbon as natural forest ecosystems.

In a number of publications (Houghton, 1997; Houghton et al., 1996; *Global Environment Outlook 2003*, 2002; Kondratyev et al., 2001a, 2002b; Kondratyev, 2002; Kondratyev and Varotsos, 2003), boreal forests and wetlands of the northern hemisphere as well as wooded tundra and tundra ecosystems are identified as the main reservoirs for CO_2 sinks on land. These ecosystems have been preserved mainly in Canada and in Russia, with the area of forests in Russia being greater by a factor of 1.5. In Russia the areas of wetlands are also great. Hence, the Canadian boreal forests, wetlands and tundra assimilate about $1,000 \, \text{Mt} \, \text{C} \, \text{yr}^{-1}$, with the share of Russia constituting $1,500 \, \text{Mt} \, \text{yr}^{-1}$. The total value of the sinks of anthropogenic carbon into natural ecosystems being equal to 2,500 Mt yr (Table 8.6). This is a maximum estimate of the sinks of anthropogenic carbon for forest ecosystems and wetlands of Russia.

The conclusion can be drawn that Russia is a territory of pure sinks of anthropogenic carbon, where its accumulation above assimilation of emissions of its own approximately ranges between 300 and $800 \, \text{Mt} \, \text{yr}^{-1}$ depending on assumed conditions of calculation. One hectare of the territory accumulates between 250 and $650 \, \text{kg} \, \text{C} \, \text{yr}^{-1}$. From the data of Mokrousov and Kudeyarov (1997), the annual natural sink of carbon in Russia varies from 750 to $5,000 \, \text{kg} \, \text{ha}^{-1}$. However, these estimates do not take into account the assimilation capabilities of the trunk (Losev et al., 2001a) and rootstock, which in the forest zone additionally provide an annual sink of carbon of more than $400 \, \text{kg} \, \text{ha}^{-1}$. Then the share of the anthropogenic sink of carbon per unit area of ecosystems constitutes 13–20% of the share of the natural sink. The biomass of boreal and temperate forests constitutes about $200 \, \text{t} \, \text{ha}^{-1}$,

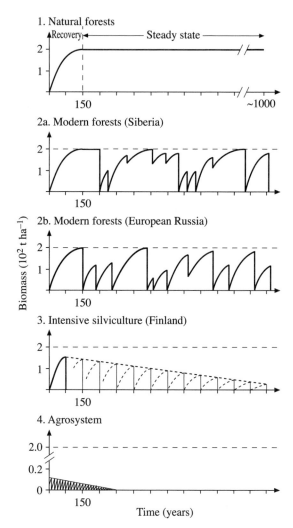

Figure 8.2 Scheme of the state of boreal forest communities at different levels of antropogenic pressure. 1. Natural state of forest ecosystems, the succession period is about 150 yr, biomass – 200 t ha^{-1}, and persistent state during thousands of years. 2a. Present forests of Siberia, with disturbances connected mainly with fires. 2b. Present forests of the European part of Russia, with disturbances connected mainly with the overall large-scale felling and, to a lesser degree, with fire. 3. Intensive silviculture (Finland), where biomass does not reach a stable state and gradually decreases. 4. Agrosystems, where biomass is very low and decreases over time (Gorshkov et al., 2000).

hence, the sink of carbon with respect to biomass averages less than one-hundredth of 1%.

   Myneni et al. (2001) made an analysis of satellite information (40,000 satellite images) with the use of the normalized differential vegetation index (NDVI) for the

Table 8.7. Sink of carbon for the forests of Canada, Russia, and the USA.

	Satellite data			Data from other sources		
Country	Pool (Gt C)	Sink (Gt C yr^{-1})	Area (Mha)	Pool (Gt C)	Sink (Gt C yr^{-1})	Area (Mha)
Canada	10.56	0.0731	239.5	11.89	0.093	244.6
					0.085	
Russia	24.39	0.2836	642.2	32.86	0.429	816.5
					0.058	763.5
USA	12.48	0.1415	215.5	13.85	0.167	217.3
					0.098	247.0
					0.020	

period from July 1981 to December 1999 to estimate the biomass increment in the northern hemisphere in temperate and boreal forests over an area of 10 million km^2. The results obtained for three countries (basic regions of carbon sink) are given in Table 8.7. As the authors noted, due to the insufficient resolution of satellite images under study, about 200 million ha of forests and wetlands have not been taken into account into Russia.

Data of Table 8.7 show that Russia is the greatest region of sinks for anthropogenic carbon. But these data are significantly underestimated for Russia due to incomplete consideration of the areas of forest ecosystems and wetlands. Also, no account was taken of sinks in other natural ecosystems (taiga and tundra). Finally, as Myneni et al. (2001) pointed out, the sink into wood biomass is only taken into account, not the sink through pure primary production. With all these factors taken into account, observational data mentioned confirm the calculations made above, but the accuracy of calculations and of empirical data is different. The error of assessment of the sink into biomass of the forests of the northern hemisphere from satellite data constitutes 50%, however, on the basis of the law of conservation (and with further assumptions) it lies nearer 20%.

There are, at least, four reservoirs of accumulation of organic carbon in land reservoirs. The first reservoir is an additional increment of wood in tree trunks. Fertilization due to the growth of concentration of CO_2 in the atmosphere also leads to the growth of the gross and pure primary production, which (relatively) rapidly goes into other reservoirs. Lakes, marshes, peat bogs, and water reservoirs are very important reservoirs of accumulation of organic carbon. Every year the lakes of the world accumulate 54 Mt of organic carbon, peat bogs 96 Mt, and water reservoirs 265 Mt giving a total of 415 Mt (Ludwig, 1997). These estimates are sufficiently conventional, and their errors are large, since calculations of the rate of accumulation of peat in the marshes substantially underestimates the annual accumulation of carbon. This is because some carbon is sunk in the solutions and suspended particles of rivers. Moreover, there are data that show that active formation of peat began not with the beginning of the Holocene, but about 5,000 years ago.

Soil is an important carbon reservoir. An additional accumulation of organic carbon in the soil with increasing concentrations of carbon dioxide in the atmosphere was measured for meadows between 1992–1997 on the test site Jasper Ridge Biological Preserve of the Stanford University in central California. It turned out that the growth of carbon concentration in the atmosphere leads to an additional accumulation of carbon (limited by the nitrogen content) in the soil (*Grassland Can Act as Carbon Sinks*, 2001). However, any quantitative estimates are absent. An additional carbon reservoir is underground water, which removes $200\,\mathrm{Mt\,C\,yr^{-1}}$.

For all land reservoirs, organic carbon is removed with water run-off into the World Ocean and partially into the closed seas and lakes. The global sink into the World Ocean is estimated at between 300 and $1,000\,\mathrm{Mt\,C\,yr^{-1}}$, with the latter estimate being more realistic, especially with the input of carbon into underground waters being taken into account (Gorshkov, 1995a, b; Ludwig, 1997). Since the Russian rivers' run-off constitutes about 12% of the world river run-off, removal of carbon from the territory of Russia by rivers can be estimated between 36 and $120\,\mathrm{Mt\,yr^{-1}}$. Thus a considerable part of the anthropogenic carbon removed from the global cycle in Russia, which accumulates in the ecosystems of boreal forests, wetlands, taiga, and tundra, cannot be estimated accurately, but the lower limit of $550\,\mathrm{Mt\,yr^{-1}}$, including about $300\,\mathrm{Mt\,yr^{-1}}$ of "foreign" carbon, seems sufficiently justified not only by calculations based on the law of conservation, but also from satellite observations of changes of the biomass of the Russian forests at the end of the 20th century (Table 8.7). *It follows from these estimates that the Kyoto Protocol should be thoroughly revised in consideration to its scientific basis and practical conclusions.*

8.3 GLOBAL WARMING AND THE KYOTO PROTOCOL

The UN Framework Convention on Climate Change (FCCC) was adopted on May 9, 1992 and came into force on March 21, 1994. At the first FCCC Conference in 1995 in Berlin it was decided to start a stagewise procedure of reducing the rate of greenhouse gas concentration growth in the atmosphere and to create an ecologo–economic mechanism to regulate these processes. This decision was realized in 1997 at the third FCCC Conference in Kyoto by adoption of the Kyoto Protocol, according to which each developed country and those with transition economies undertook to reduce greenhouse gas emissions, concentrating first of all on CO_2. By the beginning of 2001, 84 countries, including Russia, signed the Protocol. Russia signed the Protocol on March 11, 1999. From that time, many states have ratified the Protocol. For the Kyoto Protocol to come into force, it should be signed by 55 countries, including developed countries responsible for 55% of the total volume of emissions with respect to 1990 assumed as a reference point (Kondratyev, 1999a; Kondratyev and Varotsos, 2000).

Global warming and the Kyoto Protocol now became not only a field of scientific discussion but also a field of political confrontations (Demirchian et al., 2002), especially after President George W. Bush refused to sign the Kyoto Protocol. Part of the scientific community expressed doubts about various aspects of the problem of climate warming. So far, they are concentrated on climatic problems (observational network, data processing to estimate the rate of surface air temperature (SAT) increase, climate models, retrieval of paleotemperature data from the isotopic composition of ice cores, and CO_2 concentration from analysis of air bubbles in ice cores, revealing uncertainties, errors, and unreliability) and on assessment of potential economic and social consequences of a warming. Apparently, these estimates depend completely on the reliability of the climatic data and climate models (i.e., they contain many uncertainties: Boehmer-Christiansen, 2000; Kondratyev et al., 2001a).

A detailed review and analysis of the indicated disputable problems are given in the work *Global Climate Changes: Conceptual Aspects* prepared by an international group of authors (Kondratyev et al., 2001). Its main conclusion is a demonstration of the inadequacy of such (most important) international documents as the FCCC and the Kyoto Protocol. Existing observational and numerical modelling data are very uncertain and lack reliable indications of an anthropogenic signal in the rate of the SAT growth in the 20th century.

This review and other publications (e.g., Boehmer-Christiansen, 2000; Kondratyev and Varotsos, 2000; Kondratyev and Demirchian, 2001; Essex and McKitrick, 2002; Tol, 2003) reveal three groups of views of the global warming problem in the world scientific community. One group is an absolute support of the anthropogenic cause of global warming and of the Kyoto Protocol. This group comprises scientists and officials who began and continue to work in this field, "fertilized" with considerable financial investments and occupied by respective national and international structures. Another group rejects anthropogenic causes of global warming and even the warming itself. It critically assesses the results of the estimation of CO_2 concentration in the past from the analysis of air bubbles in ice cores from the boreholes in Greenland and Antarctica (e.g., Jaworowski, 1997). In some cases such views are backed by the interest of industrial and economic circles. Finally, there is a group of scientists who want to thoroughly study the global warming problem to reach an adequate understanding of its complexity and therefore reach a satisfactory scientific conclusion.

In discussions on the global warming problems, the function of natural ecosystems, which regulates and stabilizes the environment and climate, has been ignored as a rule (Kondratyev, 1999a; Gorshkov, 1995a, b; Gorshkov et al., 2000, 2001; Makarieva, 2000) as well as climatic instability inherent in the Earth, which determines a certain futility of discussions.

Both the state of the environment and climate change vary rapidly, which is usually not observed in undisturbed nature. This testifies to the physical instability of the Earth's climate (Gorshkov et al., 2000; Kondratyev, 1992). At the same time, the fact that the planet has sustained life for 4 billion years in a very narrow range of oscillations of physical characteristics (global mean temperature in particular), points to the fact that the climate was always suitable for the respective forms of life. This testifies to the existence on Earth of a specific mechanism for stabilization of the environment and climate. This mechanism is the biota of the Earth, which not only stabilizes but also forms the environment and climate (see Section 2.3). The location and character of the mean global temperature stability are shown in Gorshkov (1995a, b) and Gorshkov et al. (2000). Based on construction of the potential function (the Liapunov function) depending on mean global temperature for the range from complete glaciation to complete evaporation of the hydrosphere, two physically stable states of the planetary climate have been established – complete glaciation ("white" Earth) and complete evaporation ("hot" Earth) (Figure 2.1). The intermediate state with the presence of liquid water is unstable. To provide stability of this state, it is necessary to introduce non-physical singularities into the behaviour of albedo and greenhouse effects as functions of temperature, which can be generated by only natural undisturbed biota. Hence, it is clear that violation of the mechanism of biotic regulation and of the biogeochemical cycles of nutrients played the leading role in climate changes taking place during the period of civilized existence.

The total emission of basic nutrients – carbon from land biota due to human economic activity reached 6.3 Gt/year at the beginning of the 1990s (Section 8.2), which exceeded the emissions from fuel burning ($5.9\,\mathrm{Gt\,yr^{-1}}$). What is strikingly apparent in discussions about global warming and the Kyoto Protocol is the fact that carbon emission by the Earth's biota as a result of growing human economic activity has been completely ignored. This testifies to the senseless and ungrounded desire to establish quotas on carbon emission reduction based only on fossil fuel burning. The approach itself is incorrect: quotas on reduction of emissions from fossil fuel burning are not of importance but quotas on the share of natural ecosystems revival and preservation of existing natural ecosystems, especially forest ecosystems, which are the main areas of sinks and regulation of the carbon cycle and other nutrients on land, are of importance. Neither afforestation nor creation of forest plantations (silviculture) should be practiced. Natural forest ecosystems should be revived. Planting forests in Europe and in other global regions does not provide a revival of natural forests. It is a special kind of forestry – silviculture. This is but one more technology violating the biogeochemical cycle. On the other hand, a 30% reduction of the area of exploited forests, as estimates show (Gorshkov et al., 2000), could stop the growth of greenhouse gas concentrations without reconstruction of the present energy production (Figure 8.3).

Estimates of the sources and sinks of the main greenhouse gas (CO_2) recalculated for carbon (Sections 8.1 and 8.2) with errors taken into account, show that a 5% reduction of greenhouse gas emissions by developed countries suggested by the Kyoto Protocol (Kondratyev et al, 2001a) by 2008–2012 lies within the error of estimation of the sources and sinks of carbon.

George W. Bush justified the refusal of the USA to take part in the procedures of the Kyoto Protocol as follows: "The Kyoto Protocol is inefficient, since it excludes 80% of the world from participation in following its recommendations including the main regions of concentrated population such as China and India" (Kondratyev et al., 2001). This statement has serious flaws. Despite the US refusal to adhere to the

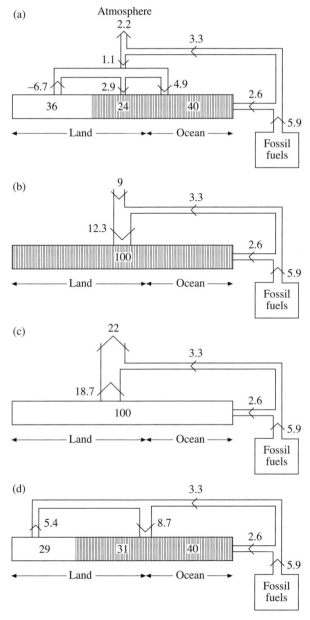

Figure 8.3 Global changes of the environment and the virgin biota. Change of the global carbon cycle: (a) – present state of the biosphere; (b) – completely undisturbed land and ocean biota; (c) – completely developed global biota; and (d) – stopped global changes in the carbon content. Shaded areas represent virgin biota with non-shaded areas representing developed biota. Numbers near the arrows denote carbon fluxes in $Gt\,C\,yr^{-1}$. Numbers in shaded and non-shaded areas denote pure primary production in $Gt\,C\,yr^{-1}$ (Gorshkov and Makarieva, 1998).

Protocol, it is much fairer to the USA than to Russia, though Russia is not recommended to reduce greenhouse gas emissions since it is the natural ecosystems preserved in its territory that serve as sinks for a considerable part of the global carbon emissions (in addition to the sinks of its own carbon emissions).

Discussions on sinks and emissions with changing land use and forestry in preparation of the Kyoto Protocol were carried out on a non-scientific basis, since only one mechanism of sink on land – planting and growth of forests – has been considered. This mechanism was discussed in an incorrect aspect: with the growth and development silviculture, emissions of carbon resulting from previous forest felling is compensated only partially (especially if, instead of natural revival, industrial forest systems are created) (Figure 8.2). An experience of forest felling in Europe has shown that planted deciduous forests have been substituted for earlier natural oak and beech forests, and their total biomass is half the biomass of the previous forests (*Protecting Tropical Forest: A High Priority Task*, 1990). Hence, they have compensated for only half the emissions due to earlier deforestation. Meanwhile, natural climacteric forests, according to the theory of biotic regulation of the environment, can regulate CO_2 concentration in the atmosphere, removing a surplus or replenishing a deficit of carbon. Estimates of this response of virgin forests are given in Section 8.2 (for Russian territories). Only the preservation and reconstruction of virgin forests (not silviculture), creates a real compensating mechanism for the sink of excess anthropogenic carbon from the atmosphere.

Was the decision of the Kyoto Protocol correct in stating that the sides have no right to claim an account of carbon sinks of exclusively natural character (Grabb et al., 2001)? From our point of view, this decision is incorrect: it is advantageous for developed countries which destroyed their natural ecosystems long ago (initially, forests and wetlands) as well as many developing countries which substituted their natural landscapes for agrosystems. At the same time, it restricts the interests of the countries, both developed and developing, whose natural ecosystems have been preserved. Replacing natural ecosystems by anthropogenic systems increases emissions of carbon (due to changing land use), a fact which is also neglected in the documents of the Kyoto Protocol. Hence, these groups of countries benefit twice. Countries with preserved virgin landscapes are not advantaged, and the decision made within the framework of the Kyoto Protocol is discriminative against them.

The total destruction of natural ecosystems in countries where practically none remain, was carried out to develop economics, to raise the living standard at a cost of a lost ecological resource, which performed a basic ecological function – regulation and stabilization of the environment – and to further develop the global ecological crisis. *Within the market economy, natural ecosystems should be considered a most important ecological resource together with mineral resources, soil, and bioresources. If any country has exhausted its ecological resource, it is apparent that an increasing burden is laid on the ecosystems of the country where this resource has been preserved. Countries therefore begin to use a "foreign" ecological resource – obtaining ecological services at the expense of foreign territories.* The World Ocean, being world property, is an exception. It is a sink for a considerable part of the world's anthropogenic emissions of carbon and pollutants, and each country has a part of this sink of its own equal to its share of the global emission (see Section 8.2). As for the sink of excess anthropogenic carbon into natural ecosystems of land, this sink should be taken into account in calculations of the balance of anthropogenic carbon for each country. If it turns out that such ecosystems in some countries absorb foreign emissions of carbon, it is necessary to coordinate an amount of compensation from the world community accounting for these additional loads on preserved ecosystems. Apart from the Russian Federation, such countries are Canada, Austria, probably Brazil and some others.

Using the same method as that used to calculate the budget of anthropogenic carbon in Russia, so the budget for the continental landmasses was also calculated, taking into account data on the industrial emissions of carbon in these regions (*Global Environment* ..., 2002) and data given in Table 8.2.1. Estimates of the size of territories with destroyed and preserved ecosystems (Hannah et al., 1994) were also used. The result was quite unexpected.

The following regions were selected: Europe (without the former Soviet Union countries), Asia (without the former Soviet Union countries), Africa, North America, South America (Latin America and the Caribbean basin), Australia, and Oceania. Calculations have shown that the greatest destabilizer of the climate system is not the USA or Europe, as usually considered, but Asia, with emissions due to industry and land use being the greatest in the world. Another large destabilizer is North America, where industrial emissions contribute most, but the contribution due to land use is also large. The third regional violator of the climate system stability is Europe, where the industrial emissions contribute the most. The rest of the regions are territories of pure sinks of anthropogenic carbon. South America contributes more than Russia. Australia and Oceania make a marked contribution, and the contribution of Africa is negligible. In the near future Africa will become a region of climate destabilization, first of all, due to active destruction of natural ecosystems. In the last two regions, land use contributes most to carbon emissions.

One of the most positive aspects of the Kyoto process is the attention given to natural ecosystems of the World Ocean and land, resulting from its shortcomings.

8.4 HOW MUCH DOES NATURE COST?

As far back as the early 20th century a lecturer from Cambridge, economist Arthur Pegou, proposed to include in the system of economic calculations the so-called externals (i.e., natural resources not included in the system of market prices). The market economy is directed toward obtaining a maximum profit from the use of given resources. The profit is calculated from market prices, and the cost of resources includes only those having a market price. Those resources, which are impossible to sell have neither price nor value. Therefore, the market economy considers everything from the viewpoint of price, and since something has no price, like, for instance, natural ecosystems, it neither has significance, nor value, nor need, etc. Thus outside the sphere of the traditional economy are those natural objects or, using the economic terminology, natural resources, which still have not become objects of market relations.

As for ecological outlay and profits, they remain outside an economic analysis with rare exceptions, such as when this is a legal requirement, by human will, or by international agreements. The Montreal Protocol, which dealt with the reduction of ozone-destroying substances, the Kyoto Protocol, and other international agreements are steps in the right direction, though all of them also have a forced element. The Kyoto Protocol differs from other international agreements in that it connects the problem of greenhouse gas reduction (mainly, carbon dioxide) with the functioning of ecosystems. Unfortunately, so far, none of the ecological international documents discuss the problem of preservation of natural ecosystems (landscapes) as regulators and stabilizers of the environment (i.e., the most natural mechanism of life maintenance on the planet). At best, the problem of biodiversity preservation is discussed, though it is apparent that without preserving natural ecosystems (not in the form of reserves and national parks) which provide a stabilization and regulation of the environment (Figure 8.3), biodiversity cannot be preserved.

The world's scientific community is gradually coming to understand the leading role of life in creation, organization, regulation, and stabilization of the environment (see Chapters 3 and 7). To some extent, this has been recognized in the Amsterdam Declaration, the resulting document of the International Conference in July 2001 in Amsterdam "Challenge to Changing Earth". Before the conference a brochure was issued (Steffen and Tyson, 2001) in which it was noted that "the Earth is a system in which life itself helps to control its state. Biological processes closely interact with physical and chemical processes in the formation of the environmental characteristics, but biology plays a much more important role in maintaining the limits of habitability of the environment than it was assumed earlier." This statement is of fundamental importance, but it would be more appropriate to quote the theory of biotic regulation of the environment developed in the 1980s and the subject of the monographs by Gorshkov (1995) and Gorshkov et al. (2000).

The role of the natural and quasinatural ecosystems of Russia in regulation of the anthropogenic carbon cycle has been demonstrated above. These ecosystems based on feedbacks open the practically closed carbon cycle and remove part of the anthropogenic carbon into the middle- and long-term reservoirs. They accumulate both Russian sources of carbon (due to industry and land use) and part of the global anthropogenic emissions. Without this important function of the natural ecosystems of Russia, the global emission could have been 300–800 Mt yr^{-1} greater.

In the process of preparation and development of the documents for the Kyoto Protocol expenditures were estimated to account for a reduction of an emitted 1 t C in developed countries. Based on these and other estimates we will try to estimate the cost of removal of 1 t C by Russian ecosystems. Of course, it will be an estimate of the cost of only one of numerous functions of the ecosystems, therefore it is only an indirect estimate of their real cost which, of course, should be higher (Kondratyev et al., 2003).

From expert estimates, the supposed expenditures on fulfillment of obligations on reduction of carbon emissions for the USA and some European countries constitute US$150–300 per $1\,t\,CO_2$ and for Japan more than US$300 per $1\,t\,CO_2$ or US$550–1,100 per $1\,t\,C$ (Bedritsky, 2000). Hence, from a minimum estimate, the ecosystems of Russia remove carbon every year worth US$160–325 billion (300 Mt is multiplied by US$550–1,100). Even with prices an order of magnitude lower (since some models assume such prices), the sum of annual "investment" from Russia constitutes US$16–32.5 billion yr^{-1}.

According to the Worldwatch Institute in Washington, to reduce carbon emissions, it is necessary to introduce a "carbon tax" and make it US$50 per $1\,t$ in order to stimulate a reduction of fossil fuel consumption, an improvement in technologies of fossil fuel burning, and resource saving. In this case a minimum cost of removal of excess carbon in Russia would be US15 billion yr^{-1}. From the estimates of this Institute, such a tax will raise the price of 1 litre of petrol by 4.5 cents, and the price of $1\,kWh$ by 2 cents.

Another way to reduce carbon emissions is the transition to alternative sources of energy. There are estimates of the cost of providing solar and hydrogen sources of energy (Kondratyev, 1999a). To put them into practice, it is necessary to introduce a tax on the use of fossil fuel (carbon tax) from US$70–660 per $1\,t$. This means that the price of excess removal of carbon by Russia's ecosystems will constitute between US$21 and 200 billion yr^{-1}.

In the USA calculations have been made of the costs of halving emissions of carbon (Bates, 1990) by substituting energy of fossil fuel with atomic energy. They have shown that for this purpose, US$50 trillion and an introduction of one nuclear energy reactor every 2.5 days over $38\,yr$ would be required. Bearing in mind that halving the emission of carbon in the USA at the moment of estimation constitutes about 700 Mt, a minimum price of only excess removal of carbon by the preserved ecosystems of Russia would constitute more than US$20 trillion. These estimates show the inefficiency of substitution of energy sources for fossil fuels in order to reduce carbon emissions.

All figures show that preserved natural ecosystems in Russia are most valuable and that their value exceeds that of all the mineral resources in Russia. From estimates (Putin, 2000), the total cost of mineral resources constitutes not less than US$28 trillion, and an estimate of their profitable part constitutes only US$1.5 trillion. Comparison of the cost of usable raw material resources with the "price" of only one function of natural ecosystems shows that at present it is much more profitable to preserve these ecosystems. If Russia could get compensation of US$150 per $1\,t$ for removal of 300 Mt yr^{-1} of foreign carbon (about US$40 per $1\,t\,CO_2$), it would amount to US$450 billion per 10 years. During this period all the foreign debt of Russia would be removed, with enormous sums being invested in the economy as well as in social and ecological projects. Apart from the excess 300 Mt, the ecosystems of Russia assimilate 250 Mt C yr^{-1} of its own (with 450 Mt of Russian emissions being assimilated by the World Ocean). This constitutes US$375 trillion for a decade. Thus Russia with scientific and diplomatic investment could gain huge profits and preserve its mineral resources on practically virgin territories. The value of these territories will increase in the future (Kondratyev et al., 2003a).

In this connection we believe that it is not the proper time for accomplishment of these recommendations of the Kyoto Protocol in Russia, excluding joint implementation and clean development. The next phase of the Kyoto Protocol should take into account the total anthropogenic emissions of carbon, including emissions due to land use. In this connection the contributions of various countries to global emission should be re-evaluated. After an assessment of the carbon sinks in the World Ocean, it is necessary to determine the countries/donors whose preserved ecosystems assimilate an additional mass of anthropogenic carbon and to establish an order of compensation for the use of their ecological space. Russia should draw up an inventory of its total greenhouse gas emissions and specify the initial sum of its global contribution for 1990 (*Development of mechanisms for the Greenhouse Gases Emission Quota Market*, 2001). Last but not least, the problem of the contribution of anthropogenic carbon to the observed global warming remains unclear since, most likely, the main contribution is made by changes of the water vapour content in the atmosphere and cloudiness as well as changes of the global albedo due to ecosystem destruction.

Of critical importance is the understanding by the world community that it is not only in the "price" of natural ecosystems and landscapes, but also their determining role in the preservation of the stability of the environment. Destruction of the controlling natural mechanism leads to breaking the stability of life on Earth on all scales, from molecular to global. In this light, it is necessary to reorient scientific studies and international programs, especially the International Geosphere–Biosphere Programme (IGBP), towards studies of the regulating functions of ecosystems.

As far as the Kyoto Protocol is concerned the President of Russia Vladimir Putin said in his speach at the World Conference on Global Climate in Moscow (29 August 2003): "The government is thoroughly considering and studying this issue as well as the entire complex of difficult problems linked with it. The decision will be made after this work has been completed. And, of course, it will take into account the national interests of the Russian Federation."

9

Territorial expansion as one of the mechanisms for enhancement of damage from natural hazards

9.1 INCREASE OF NATURAL HAZARDS (NATURE RISKS) AND DAMAGE RESULTING FROM THEM

With the development and complication of our present civilization, the growth of the size and density of the population on the Earth and, as a consequence, the expansion of developed territories, has increased the number and scales of technogenic catastrophes (Kondratyev et al., 2002a). Nature also responds with a growth of catastrophes, which seriously damage the economic and social development of humankind. Damage from natural and technogenic catastrophes is well known. From the data of the German insurance company "Munich Re" (*Global Environment Outlook 2003*, 2002), the 1990s were record years for amounts of natural hazards leading to losses of tens of thousands of lives (Figure 9.1). In 1999, 700 natural catastrophes took the lives of more than 70,000 people. The economic losses from natural cataclysms for the last 10 years constituted US$535 billion, of which US$160 dollars were attributed to 1996 and almost US$80 billion to 1995.

In the developing countries where the population and territorial expansion grows especially rapidly, relative losses in gross domestic product (GDP) due to natural catastrophes are much greater than in the developed countries, and the absolute amount of victims is always higher. From 1964 to 1991, in Africa, for instance, 139,262,690 people were injured as a result of natural hazards and 715,320 people were killed (Mohammed and Abdel Rahman, 1998). This is, respectively, 22% and 1.1% of Africa's total population, and thus almost a quarter of the population has been affected.

The book *State of the World 1999* (2000) discusses natural hazards in the world in 1998, which, in terms of losses of lives, was just above average for the last decade of the 20th century: "In 1998 we witnessed the most devastating floods during the whole history. China suffered very much: by expert estimates, the damage from the flood in the regions of the Yangtze River was calculated at US$36 billion.

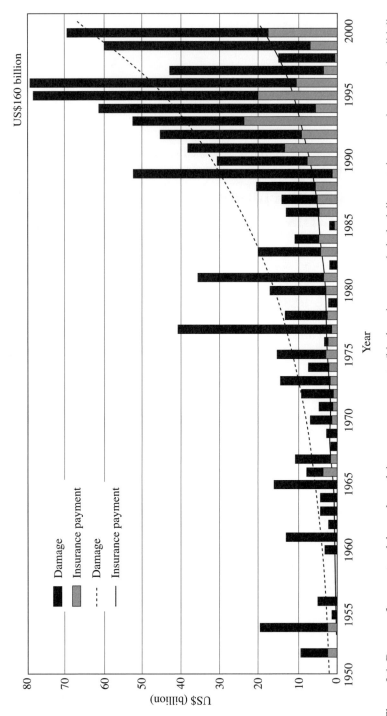

Figure 9.1 Damage from natural hazards and insurance payments (black columns and dashed line, and grey columns and solid line, respectively). Damage was especially great in the 1990s, with maximum in 1995 and 1996.

After the flood, the most severe during the last 44 years, 2,500 people were lost and 56 million people remained homeless.

Bangladesh has endured a very long season of rains and as a result, about two-thirds of its territory remained under water for an entire month, with 21 million people becoming homeless. This flood, the severest in the history of the country, destroyed part of the yield of rice and forced some owners of textile factories in Dacca, the capital, to close them for several weeks, depriving the country of the two most important sources of revenue from export. In 1998, 58 countries suffered from floods.

Clearly, hurricane Mitch widely discussed in the press is a vivid example of what the result of global warming can be: a storm of record force fell on Central America, with wind speeds reaching $270 \, \mathrm{km \, hr}^{-1}$. Its northward movement turned out to be blocked by a powerful atmospheric front, and the trapped hurricane deposited a 2-m layer of water on some regions of Central America over a period of 1 week. Water flows were so powerful that they destroyed houses, schools, and factories, and destroyed roads and bridges.

Honduras and Nicaragua suffered the most. The event has led to huge geological changes. From available data, about 70% of the yields in Honduras were washed away with water flows, shifting thick upper layers of soil which formed more than 1,000 years ago. Destructive mudflows have changed the topography of the landscape and wiped out dozens of villages, in some cases burying buildings and people. About one-third of the population of Honduras remain homeless. Preliminary data for these two countries show that 11,000 people were lost and about the same amount disappeared. President Flores of Honduras, announced that "in total, during several days everything created during 50 years was wiped out." There were many who said that after such a catastrophe the economics of Honduras and Nicaragua would not be able to recover completely for many years.

The destabilized climate can lead to strong and prolonged heat. In Texas during such a hot period, 100 people died when the air temperature in Dallas over many weeks did not drop below 35°C. In India, 3,500 people died from heat, unheard of for 50 years. In 1998, 45 countries suffered from severe drought, which caused forest fires in many of them.

Tropical rain forests usually do not burn, but drought has led to unprecedented fires. In South-East Asia they began at the end of 1997 and in the region of the Amazon River fires raged for the whole year in 1998. In the spring of 1998, fires began in the south of Mexico causing an alarming situation in the air quality from this region right up to Chicago. In the early summer, forest fires reached the sub-tropical forests of Florida, which necessitated evacuation of an entire county. In Russia, a combination of severe heat, drought, and a poorly controlled economy has led to a decrease in grain yield to its lowest level for 40 years.

Natural hazards, in contrast to technogenic catastrophes, are more rare but they have a powerful destructive effect and often cover large territories. The developing countries suffer from them most often. They have neither sufficient information nor methods to predict such events, a lack of estimates and methods to control these risks, poor governmental structures to respond to and resolve the consequences of natural catastrophes, and poor knowledge of insurance and legal spheres useful to solve such problems.

The conomic consequences of natural hazards depend on the type of natural process, the size of the damaged territories, the structure of economic development and present economic situation in the region, and, finally, population density.

Different causes of the origin of dangerous natural processes taking the form of catastrophes and leading to emergency situations can be identified. First of all, these are natural causes connected with the geological and geophysical processes in the Earth's crust as well as processes in the atmosphere and in the World Ocean. They are of stochastic character with aperiodic changes of their frequency of occurrence and intensity. The effect of cosmic factors (solar activity, parade of planets, etc.) is actively discussed. Therefore, it is difficult or even impossible to predict dangerous natural processes, but methods of their reference to certain territories can often be found.

Other causes are social and economic:

1   The growing economic expansion leading to changes of the environment due to destruction of natural ecosystems (landscapes). This changes the Earth's albedo, the continental water cycle, and greenhouse gas concentrations in the atmosphere, leading to climatic instability on the planet. One of the consequences is the growth of the number of extreme events, part of which turns out to be natural catastrophes.

2   The growth of the density and size of population in the developed territories. Estimates obtained for the USA have shown that during the period 1932–1997 the growth of losses from floods constituted from US$1 billion in the 1940s to US$5 billion in the 1990s. Forty-three percent of the growth of losses are connected with the growing size and density of population, and the remaining 57% are mainly determined by the growth in well-being of the population and, to a lesser extent, by increasing precipitation and consequent floods (Anatta, 2001).

3   The development of new territories in the 20th century considered earlier to be almost unsuitable for life, mainly because of the increased probability of the development of dangerous natural processes. Such territories either have not been studied preliminarily or have been investigated inadequately. Therefore, they often become territories of increased danger.

4   The development of a specific complex of dangerous natural–technogenic processes provoked by human activity (waterlogging of territories, erosion of the shores of water reservoirs, directed seismicity, landslides, etc.). The cause lies in the low predictability of the development of natural processes and with the increased anthropogenic interference into natural processes.

5   Errors in the strategy of the prophylactic safety insurance of the regions. As a result, all forces and finances are spent on repairing the consequences of catastrophes not preventing them.

Evaluating the regions of Russia based on analysis of natural hazards, the

Ministry of Emergency of the Russian Federation notes that the highest potential for the development of natural emergencies in the near future are in the Yakut Republic, Chukot Autonomous Okrug, Kamchatka Oblast, Murmansk Oblast, Kalmyk Republic, North Osetia, Adygea, Ingushetia, Karachaevo-Cherkessia, Chechen Republic, Dagestan, Primorski Krai, Amur Oblast, and Sakhalin Oblast.

Dangerous natural events in Russia and their consequences

For a long time, Russia has been intensively developing due to gaining mastery over new territories and new resources, even over those territories where there is a likelihood of increased frequency of occurrence of dangerous natural processes. At present the average multiyear damage from dangerous natural processes in Russia constitutes 7–10% of GDP each year. It should be noted that damage connected with social losses (expenditures on recovering health, reconstruction of dwellings, work, and the psychological cost of despair) is excluded from this estimate, although those are the main criteria of human life. Also, no account is taken of the losses connected with economic activity within the damaged region, its budget, and investment policy.

Two problems remain unsolved in the studies of natural risks:

1   At present it is impossible to predict precisely when and where a catastrophe will occur and what damage will be done. So it is difficult to speak about any substantiated cost of damage.
2   Insufficient statistics on the geology–geographic propagation of the natural risk over many regions and, hence, the impossibility of a reliable determination of the objects for which the natural danger is greatest.

To solve these problems, first of all, it is necessary to dwell upon the existing ideas of natural disasters. A natural disaster is defined as a suddenly occurring local ecological situation with a harmful impact (Efimov, 1996). However, there are natural disasters, which develop slowly and act over long periods, sometimes for many years. For instance, the drought in the Sahel (Africa) continued for several years. The process of desertification can be called a slow-developing or prolonged natural disaster.

A natural disaster always combines three factors: an extreme geophysical event (the centre of the catastrophe), an impact on land surface (factor of damage), and an inability of the population to withstand the event (vulnerability). To minimize the damage from a natural disaster, one should learn to influence these three factors. This influence is provided scientifically through prediction, but more often through scenarios, protection, and prevention of a natural disaster. Most natural disasters are characterized by relatively uncertain times of occurrence and different consequences which, in turn, depend substantially on factors such as the geographic location of the region, its geological structure, the level of economic and social development, the size and density of the population, its living conditions, and cultural level.

Most urgent is the problem of reducing the danger of natural disasters. Many countries have accumulated data sets on natural disasters and their increase. Analysis carried out by the Yokohama Conference on Sustainable Development (1994) has shown that most widely spread natural processes with the heaviest consequences are earthquakes, tsunamis, volcanic eruptions, snow avalanches, mudflows, landslides, floods, droughts, frosts, dry winds, storms, hurricanes, cyclones, and typhoons (Osipov, 1995, 1997; Kondratyev et al., 2002a). All these events have been compared in three categories: amount of damage, amount of victims, and amount of losses. Conclusions formulated in the materials of this Yokohama Conference are unconsoling. On the whole, the number of natural catastrophes in the world is constantly increasing. Especially frequent are catastrophes whose damage constitutes 1% or more of GDP of the country. Most dangerous and frequent natural events are floods constituting 32% of the total number of events. These are followed by tropical storms (30%) and earthquakes (22%).

In the present insurance literature, natural disasters are classified according to three indicators: cause of occurrence, propagation velocity, and possibility to limit the area of damage.

The causes of occurrence are divided into three groups: natural forces (earthquakes, tsunami, etc.), biological factors (epidemics, epizooties, epiphytoxics), and technogenic factor (accidents at industrial enterprises provoking dangerous natural events).

Propagation velocity is divided into: suddenly propagating (earthquakes, avalanches, landslides, etc.) and gradually propagating (hurricanes, floods, etc.).

The possibility to limit the area of damage for natural disasters is subdivided into non-localizable (earthquakes, hurricanes, etc.) and localizable (avalanches, landslips, etc.).

Most natural disasters cannot be prevented completely. But, for instance, with the help of insurance, their consequences and impact can be reduced to a certain extent, since insurance often requires preventive measures.

There are three periods in the struggle against natural disasters: periods of warning, localization, and repair of consequences.

In the battle against natural disasters, a system of warnings and emergency action situations has been in place in Russia since 1992. At the international level a project was worked out at the "Convention on the Coordinated Aid in Case of Emergency of Natural and Technogenic Character". Many natural processes were studied, as were trends in their development and propagation. Accumulation, systematization, and analysis of data of manifestations of natural processes enable one to assess the danger for each region and the appropriate admissible level of risk (i.e., from the viewpoints of insurance companies).

In the territory of the Russian Federation, on average, up to 250 extreme events connected with dangerous natural processes occur every year. Koff and Chesnokova (1997) performed an analysis of the multiyear-mean social–economic damage from most widespread natural (meteorological and geological) processes over the territory of Russia. The Ministry of Emergency Situations data and an enormous amount of actual material accumulated at the Institute of Lithosphere of Bordering and Inland Seas, have made it possible to assess damage from 12 natural processes over all regions of the Russian Federation. The scenario of the development of dangerous natural processes over the territory of Russia shows that the future of the country will not be calm. Therefore, despite the state of the economics, scientific investigations are needed to develop theoretical and methodological bases of assessment of the territories subject to dangerous natural processes, to analyse the vulnerability of both the country and people, and to create new machinery and technologies used for protection.

9.2 PROLONGED NATURAL DISASTERS ON REGIONAL AND GLOBAL SCALES

The growth in number of natural disasters and damage from them is closely connected with human expansion into new territories and the increasing density of population in developed and developing lands. Damage also grows with an increase in human wealth. Natural disasters are brought about by the destruction of natural regulators – ecosystems – that provide environmental stability. A marked increase in natural disasters occurred at the end of the 20th century (after 1985): between 1986 and 2000 the total number of floods, earthquakes, storms and other natural disasters tripled compared with the 1960s and insurance losses increased fivefold (Figure 9.1). These are verifiable data as these disasters mainly happened in developed countries.

The 20th century, especially the latter half, was full of natural disasters, many of which were not sudden but gradual. For a long time they manifested on a local scale, but in the second-half of the century they acquired regional and even global scales. The human activity connected with changes of the face of the planet triggers these slow-developing or prolonged dangerous natural events.

Humankind cannot predict the consequences of his intrusions into natural processes, which are much more complicated than his ideas of them. Natural processes and phenomena have a limit (horizon) of predictability, and with their violation due to economic activity, possibilities of prediction are reduced.

Another cause of such human activity lies in the risk inherent in humans, the extent of which is different for different ethnic cultures but inherent in all ethnoses (Miagkov, 2001). Finally, faith in science and technology is too strong, bringing forth the so-called technological optimism – the assurance that any problems can be solved with the help of science and technology. Meanwhile, even technological optimists understand the necessity to obtain estimates of the impact on the environment when projecting new objects and working out a procedure of environmental impact assessment (EIA). The EIA methods are still very primitive and only give some idea of the possible changes in the environment and of how natural disasters can be provoked by economic activity. They will never allow us to predict future consequences.

There are many examples of natural risks caused by human activity at local levels. An example is the destruction of the shores of water reservoirs, usually occurring after filling. They follow non-predictable rules. They can take the form of induced earthquakes in the regions of large water reservoirs or oil fields and initiation of landslides from overloaded construction works, etc.

Another example of prolonged disaster is the effect the introduction of rabbits had in Australia, which in the absence of predators became a scourge for vegetation and agriculture. The problem of introduction (deliberate introduction of exotic species of organisms into ecosystems) and invasion (non-deliberate introduction of exotic species into a given ecosystem) have become an example of a global prolonged natural disaster.

Acidification of soils and water bodies is another example of a prolonged regional natural risk. As a result of the increased fossil fuel burning and increased emissions into the atmosphere of sulphur dioxide and nitrogen oxides, acid rains fall onto territories far from the place of emission. Finally, as a result of a long chain of cause-and-effect feedbacks, lakes and rivers obtain dissolved compounds of aluminium fatal for aquatic organisms, practically "sterilizing" the water bodies and in other regions causing disease in forests.

An example of the appearance of global natural risks due to economic activity is the regular epidemics and pandemics of various kinds of influenza originating at piggeries and duck farms in China, where the multiple transfer of the influenza virus among birds, pigs, and humans led to the appearance of new forms and to their propagation over the globe. The growth in density and size of population and the ease of travel between countries played a role, too. Plague in the Old World became global in scale when global population reached 400 million, due to its rapid growth in Europe (Daily and Ehrlich, 1996). The size and density of the population marked the threshold that favoured rapid propagation of the disease. The first pandemic of influenza happened at the beginning of the 20th century with the population at about 1,600 million. Finally, AIDS became a pandemic when the size of the global population reached about 5,000 million. Nobody can predict what may happen beyond the threshold of 6, 7, or 8 billion people.

However, the most powerful jolt to the development of natural hazards has been the destruction of natural land ecosystems due to economic activity, which has violated the mechanism for regulation and stabilization of the environment and climate. Now this is universally recognized and verified by observations of the global environmental changes. The Amsterdam Declaration (Kondratyev and Losev, 2002) states:

The dynamics of the system "Earth" is characterized by the presence of critical threshold values and sudden changes. A human can unintentionally cause the processes, which will seriously affect the environment and the population. During the last half million years the system "Earth" functioned in various conditions, sometimes with sudden transitions (continuing for decades and less) which could be less suitable for humans and other forms of life. The probability of a sudden change of the environment due to human activity has been estimated quantitatively, but it should be taken into account. As for some key ecological parameters, the system "Earth" has gone far beyond the limits of variability taking place at least during the last half million years. The nature of all present changes occurring in the system "Earth", their scale and rate are unprecedented. At present the Earth is functioning in a state without analogues in the past.

Unfortunately, with the correct statement of the existence in nature of critical thresholds, despite apparent changes, the Amsterdam Declaration does not mention the fact of exceeding the main threshold – carrying (economic) capacity – of the permissible disturbance of the biosphere, which has triggered the "dominoes" effect in the environment. From the beginning of the 20th century, when the limit of stability was exceeded, uncontrolled changes have begun in the chemical composition of atmospheric chemistry, natural freshwater, and soils. The process of disappearance of species of organisms on the planet suddenly accelerated, the climate system became less stable. All these phenomena called "global ecological problems" or "global changes" are the global prolonged dangerous natural disasters, which gradually lead the system (Earth) from the threshold of transition to an unstable state and to a catastrophic threshold. The process of destruction by the present civilization of natural ecosystems (i.e., the mechanism regulating and stabilizing the space of life on the planet) is ongoing.

Ecosystems preserved on 37% of land and the still non-destroyed ecosystem of the World Ocean slow down the process of transition to catastrophe, which was shown in Chapter 8 with the CO_2 cycle as an example. The growth of the amount of numerous local and regional dangerous natural events is connected with this cardinal violation of the environment. As a result, humankind becomes less protected in this "armor of civilization" which it has built round itself. It means that numerous natural processes leading to catastrophic consequences are caused by both deliberate and non-deliberate chaotic impacts of humans on nature. Deliberate actions are connected with the modernistic system of views, which permits any impact of humans on nature and obtaining short-term profits measured by the time interval not exceeding human life spans, without taking into account the needs of the next generations. Non-deliberate actions are determined by human ignorance of the laws of the biosphere and limits of civilization expansion following from these laws.

The Amsterdam Declaration reads in this respect:

There is an urgent need to determine the ethic frames for the global government and strategies of this government of the system "Earth", since the accelerating transformation by man of the Earth's environment is unstable. Hence, a continuation of economic activity in the former (present) regime by the principle "business as usual" cannot be applied to the case of the system "Earth". Instead, considered strategies of an effective government are urgently needed to maintain the Earth's environment with a simultaneous solution of the problems of social development.

Here an insufficient understanding of the complexity, and at the same time simplicity, of the problem is observed. The present human activity is not control but destruction of the environment (see Chapter 7). If in the future humankind moves away from the "business as usual" approach which is destructive for civilization (so far there are few signs of this happening), it will still have to learn not to try and govern the system "Earth" (i.e., the biota and their environment, see Figure 7.1). This is a very complicated system and it is impossible to govern (Gorshkov et al., 2002). At the same time, the problem is clear – it is necessary to return below the banned threshold, which has been exceeded. But for this purpose, humankind and each individual should learn to control himself, rejecting, first of all, the stereotypes and myths of modernism (see Section 7). The biosphere will do everything else itself. In this case, ethic frames are really needed and new values, other than those of modernism and information society. An ecological imperative should be at the centre of the new ethics, which does not prohibit the use of nature for the benefit of mankind, for its prosperity and development, but limits its economic expansion, following from the law of the biosphere, in particular, from the law of energy distribution by the size of consumers (Figure 3.1). *The economic expansion and the growth (but not development) in all directions – are simply following the genetic programmes of humans as a biological species. Such an "instinctive" evolution of civilization has led it to confrontation with nature, which can only end in collapse and, probably, disappearance of the species* Homo sapiens.

Conclusion

It follows from the history of the formation of the present civilization and its relationship with nature that its development is based not on intellect but on genetic programmes of humankind as a biological species – it is a boundless expansion, transformation from a minor destroyer of ecosystems' stability into the main destroyer, with a growth of competitive ability compared to all other biological species. All this takes place due to the use of an additional source of energy, besides solar energy. The continuous expansion has acquired global scales and is accompanied by a rapid destruction of nature, violation of biochemical cycles, and impacts of nature on humankind (including the form of natural dangers). These processes and their intensity have continued to grow. Before the second-half of the 20th century these relationships had been ignored, since during a human life time (the basic measure of time for humans) they had not been observed. At the end of the century the environment began to change over a period of one generation. People began to feel these changes. Systems have appeared to monitor environmental parameters, which confirm an unprecedented character of changes. Therefore, the illusion of environmental stability should be rejected. Many people have done this, however, a greater amount of individuals do not agree with it, possibly because they simply do not think of this problem and continue to act based on genetic programmes, not intellect. The main trouble here (if not a tragedy) is that there are many people who are directly or indirectly responsible for the increasing process of destruction of the environment and of its basis – natural ecosystems. These are the governing elected elite, bureaucracy, and corporative structures in economics. Horizons of understanding the problems for the first group do not exceed two–three terms of tenure of office. For the officials – it is length of service. In corporative structures, where everything is determined by profit, it is the term of recoupment of investments with subsequent obtaining of profit. Therefore, they do not notice (or do not want to notice) those enormous changes taking place in the environment and, even more important, do not want to understand the main cause which scientific knowledge can give them. Such behaviour is no more than a realization of human genetic programmes (aspiration for positive emotions, for expansion, and competition). Traditionally, everything is explained by technogenic pollution (and it is apparent that this is an important problem) and hopes of its reduction by technological means, remaining within the framework of the stereotype of technological optimism. This direction of civilization development leads to a crash.

Meanwhile, the environmental stability is maintained by functioning of the structured natural biota of the Earth, which does not change on temporal scales of evolution. A limitation of the impact on natural biota, which has been considerably exceeded, is determined by the requirements of the scientific laws of the biosphere. Only by obeying these laws is it possible to provide life stability as a whole for humankind. These are measures on the basis of intellect, not instinct. For this purpose it is necessary to work out the scientific basis of life stability, which requires a study of the laws of the functioning of ecosystems and the biosphere as a whole. Humankind can develop in any way in conditions of environmental stability, but only within the carrying (economic) capacity of the Earth, which it had done before the 20th century, before it exceeded the threshold of admissible disturbance to the biosphere. The scientific basis of life stability should become the main object of study and education in the 21st century.

The importance of such studies is now widely recognized. In this regard the Amsterdam Declaration (Kondratyev and Losev, 2002) notes the following:

> It is necessary to create a new system of global ecological science. It starts developing as a result of the use of the mutually supplementing approaches within the programs of study of the global changes and needs support and further development. This system will be based on the existing and expanding disciplinary basis of the science on global changes. All disciplines, problems of the environment and development, natural and social sciences will be combined. International cooperation will be based on a common reliable infrastructure, efforts directed at complete involvement of scientists of the developing countries will be intensified. To create an efficient international system of global ecological science, mutually supplementing efforts of the countries and regions will be used.

The basis for the "global ecological science" already exists. It is the theory of biotic regulation of the environment. It combines various disciplines, and it should serve as the basis for social sciences in solving the problems of civilization development.

References

Advancing Sustainable Development. The World Bank and Agenda 21 since Rio Earth Summit (81 pp.). (1997) Washington, DC: World Bank.

Aizatullin, T. A., Lebedev, V. L., and Khailov, K. M. (1984) *Ocean: Fronts of Dispersion, Life* (192 pp.). Leningrad: Gidrometeoizdat [in Russian].

Anatta (2000/2001) Population and wealth, more than climate, drive soaring flood costs. *UCAR Quart.* Winter, 9.

Armand, D. L. (1975) *Landscape Science* (286 pp.). Moscow: Mysl' Publishing [in Russian].

Arnold, D. (1999) *The Problem of Nature: Environment, Culture and European Expansion.* Madlen: Blackwell Publishing.

Arsky, Yu. M., Danilov-Danilyan, V. I., Zalikhanov, M. Ch., Kondratyev, K. Ya., Kotliakov, V. M., and Losev, K. S. (1997) *Ecological Problems: What Happens, Who is to Blame, and What is to Be Done?* (330 pp.). Moscow: MNEPU [in Russian].

Barkov, A. S. (1954) *Dictionary–Reference Book on Physical Geography* (308 pp.). Moscow: Minprosvet Publishing [in Russian].

Bates, A. K. (1990) *Climate In Crisis* (228 pp.). Summertown, CO: Book Publishing.

Bedritsky, A. I. (2000) Key problems of discussions on Kyoto Protocol. In: *Ecological Aspects of Energy Strategy as a Factor of Sustainable Development* (pp. 72–78). Moscow: Noosphere Publishing [in Russian].

Bernal, J. (1956) *Science and History of Ecology* (736 pp.). Moscow: Foreign Liter. Publishing [in Russian].

Boehmer-Christiansen, S. (2000) Who and in what way determines the global climate change policy? *Izvestiya Russian Geograph. Soc.*, **132**(3), 6–22 [in Russian].

Bondarev, L. G. (1998) *Paleoecology and History of Ecology* (106 pp.). Moscow: MGU Publishing [in Russian].

Bondarev, L. G. (1999) *History of Land Use* (96pp.). Moscow: MGU Publishing [in Russian].

Botta, A., Ramankutty, N., Foley, J. A. (2002) Long-term variations of carbon fluxes over the Amazon basin. *Geophys. Res. Lett.*, **29**(9), 33/1–33/4.

Bringezu, S., Behgrensmeier, R., and Schuetz, H. (1996) Material flow accounts including the environmental pressure of the various sectors of the economy. *International Symposium*

*on International Environmental and Economic Accounting in Theory and Practice,
5–8 March, Tokyo.*

Bryant, D., Nielsen, D., and Tangley, L. (1997) *The Last Frontier Forests* (42 pp.). World
Resources Institute.

Caring for the Earth. A Strategy for Sustainable Living (1991) IUCN, UNEP, WWF (228 pp.).

Colborn. T., Dumanoski, D., and Myers, J. P. (1996) *Our Stolen Future* (306 pp.). New York:
Dutton.

Daily, G. C. and Ehrlich, P. R. (1996) Global change and human susceptibility to disease.
Annual. Rev. Energy Environ, **21**, 125–44.

Dalaker, J. and Proctor, B. D. (2000) Poverty in the United States. *Current Population Report*.
60(210) i–xix, 1–33.

Danilov-Danilyan, V. I., and Losev, K. S. (2000) *Ecological Challenge and Sustainable
Development* (416 pp.). Moscow: Express-Tradition Publishing [in Russian].

Danilov-Danilyan, V. I., Gorshkov, V. G., Arsky, Yu. M., and Losev, K. S. (1994) *The
Environment Between Past and Future: The World and Russia (An Experience from the
Ecologo-Economic Analysis)* (134 pp.). Moscow: VINITI Publishing [in Russian].

Dediu, I. I. (1989) *Ecological Encyclopaedian Dictionary* (408 pp.). Moscow: Kishinev
Moldavian Soviet Encyclopaedia [in Russian].

Demirchian, K. S., Demirchian, K. K., Danilevich. Ya. B., and Kondratyev, K. Ya. (2002)
Global warming, energy, and geopolicy. *Proc. Russian Acad. Sci., Ser. Energy*, **3**, 18–46
(in Russian).

Development and Environment (1992) New York: Oxford University Press (308 pp.).

Development of Mechanisms for the Greenhouse Gases Emissions Quota Market, Final Report
(2001) Moscow: Bureau of Economic Analysis (pp. iv, 141) [in Russian].

Dobrovolsky G. V. (ed.) (1999) *The Structural-Functional Role of Soil in the Biosphere* (278
pp.). Moscow: GEOS [in Russian].

Dokuchayev (1948) *Doctrine about Natural Zones* (62 pp.). Moscow: Geographgis [in
Russian].

Dorset, J. (1968) *Before Nature Dies* (416 pp.). Moscow: Progress Publishing [in Russian].

Dronin, N. M. (1999) *Evolution of the Landscape Concept in the Russian and Soviet Physical
Geography* (232 pp.). Moscow: GEOS [in Russian].

Efimov, S. L. (1996) *Encyclopaedia Dictionary. Economy and Insurance* (528 pp.). Moscow:
Tserikh-PEL [in Russian].

Encyclopaedia Dictionary of Geographical Terms (1968) Moscow: Soviet Encyclopaedia.

Essex, C. and McKitrick, R. (2002) *Taken by Storm. The Troubled Science, Policy, and Politics
of Global Warming* (320 pp.). Toronto: Key Porter Books.

Fedorov, L. G. (1999) *Management of Wastes in Large Cities and Agglomerate Systems of
Settlements* (112 pp.). Moscow: Prima-Press-M [in Russian].

Gerasimov, I. P. (1985) *Ecological Problems of the Past, Present and Future of the World
Geography* (247 pp.). Moscow: Nauka Publishing [in Russian].

Global Environment Outlook (2002) London: Earthscan Publishing (446 pp.).

Global Environment Outlook 2000 (1999) London: Earthscan Publishing (398 pp.).

Gorshkov, V. G. (1995) *Physical and Biological Basis of Life Stability. Man, Biota,
Environment* (340 pp.). Berlin: Springer-Verlag.

Gorshkov, V. G. and Makaryeva, A. M. (1997) Dependence of heterozygote property on the
mammals' body mass. *Doklady RAN.*, **N3**, 428–432 [in Russian].

Gorshkov, V. G. and Makaryeva, A. M. (1998) Biotic regulation of the environment: Grounds
for preservation and restoration of natural biota over territories of continental scales.
Proceedings of the International Symposium on Biotic Regulation of the Environment.

Petrozavodsk, Russia, 12–16 October 1998 (pp. 3–20). Gatchina, Russia: PIYaF [in Russian].

Gorshkov, V. G., Kondratyev, K. Ya., and Losev K. S. (1996) If the wisdom of mother-nature is taken as an ally. *Herald Russian Acad. Sci.*, **N1**, 802–806 [in Russian].

Gorshkov, V. G., Kondratyev, K. Ya., and Losev, K. S. (1998) Global ecodynamics and sustainable development: natural-scientific aspects and human dimension. *Ecology*, **N3**, 163–170 [in Russian].

Gorshkov, V. V., Gorshkov, V. G., Danilov-Danilyan, V. I., Losev, K. S., and Makaryeva, A. M. (1999) Biotic regulation of the environment. *Ecology*, **N2**, 104–118 [in Russian].

Gorshkov, V. G., Gorshkov, V. V., and Makaryeva, A. M. (2000) *Biotic Regulation of the Environment: Key Issue of Global Change* (367 pp.). Chichester: Springer–Praxis.

Gorshkov, V., Makaryeva, A., Mackey, B., and Gorshkov, V. (2001) Biological Theory and Global Change Science. *Global Change News Letter*. **48**, 11-14.

Gorshkov, V. V., Gorshkov, V. G., Danilov-Danilyan, V. I., Losev, K. S., and Makaryeva, A. M. (2002) Information in living and non-living nature. *Ecology*, **N3**, 163–169 [in Russian].

Grabb, M., Vrolik, L., and Brek, D. (2001) *Kyoto Protocol: Analysis and Interpretation* (303 pp.). Moscow: Nauka Publishing [in Russian].

Grasslands can act as carbon sinks (2000/2001) *UCAR Quarterly*, winter, 9.

Gray, V. (2002) *The Greenhouse Delusion. A Critique of Climate Change 2001* (95 pp.). Multi-Science Publishing.

Great Soviet Encyclopaedia (1952) Moscow: Great Soviet Encyclopaedia (2nd edn) [in Russian].

Great Soviet Encyclopaedia (1970) Moscow: Great Soviet Encyclopaedia (3rd edn) [in Russian].

Grigoryev, A. A. (1932) *Subject and Problems of Physical Geography* (General principles of studying the structure of physico-geographical process at the methodologic front of geography and economic geography, pp. 45–49). Moscow: Sotsekgiz [in Russian].

Grigoryev, A. A. (1934) Problems of dynamic physical geography. *Proceedings of the First All-Union Geographic Congress* (Issue 2, pp. 65–76). Leningrad: Geografgiz [in Russian].

Grigoryev, A. A. (1936) On chemical geography. In: *To Academician V. I. Vernadsky on the Occasion of the 50th Anniversary of His Scientific and Educational Work*. Moscow: USSR Academy Science Publishing [in Russian].

Grigoryev, A. A. (1943) The law of the physico–geographical process intensity. *Izv. VGO*, **75**(1), 3–13 [in Russian].

Grigoryev, A. A. (1965) *Development of Theoretical Problems of the Soviet Physical Geography (1917–1934)* (246 pp.). Moscow: Nauka Publishing [in Russian].

Gromov, V. I. (1965) *Brief Review of Quaternary Mammals*. Moscow: Nauka [in Russian].

Hannah, L., Lohse, D., Hutchinson, Ch., Carr, J. L., and Lankerant, A. (1994) A preliminary inventory of human disturbance of world ecosystems. *Ambio*, **23**(4–5), 246–250.

Holdgate, M. V. (1994) Ecology, development and global policy. *J. Appl. Ecol.*, **31**(2), 201–211.

Houghton, R. A. (1997) Terrestrial carbon storage: Global lessons for Amazonian research. *Ciência e Cultura*, **49**(12), 58–72.

Houghton, R. A., Meira Filho, L. G., Callender, B. A., Hariis, N., Kattenberg, A., and Maskels, K. (eds) (1996) *Climate Change 1995. The Science of Climate Change* (968 pp.). Cambridge: Cambridge University Press.

ISO 617-3. (1993) Geneva: International Standards Organization.

Ivanov, A. V. (1999) Conceptual approaches to revealing the structural-functional organization of the biosphere. *Biosphere and Its Noospheric Way of Development* (pp. 9–61). Khabarovsk-Birobijan: IVEP DVO RAN [in Russian].

Isachenko, A. G. (2002) Is there a necessity to revise the fundamental notions in landscape science? (Comments on the paper by Kondratyev, K. Ya., Losev, K. S., Ananicheva, M. D. and Chesnokova, I. V.) *Izv. Russ. Geogr. Soc.*, **1**, 37–41 [in Russian].

Isaev, A. S. and Korovin, G. N. (1999) Carbon in the forests of the Northern Eurasia. In: *Carbon Cycle in the Territory of Russia* (pp. 63–95). Moscow [in Russian].

Jaworowski, Z. (1997) Another global warming fraud exposed: Ice core data show no carbon dioxide increase. *21st Sience and Technology*, **10**(1), 42–52.

Jensen, M. M. (2003) Consensus on ecological impacts remains elusive. *Science*, **299**(5603), 38.

Kapitsa, S. P. (1995) A model of the growth of population on the Earth. *Progress of Physical Sciences*, **26**(3), 111–128 [in Russian].

Keeling, R. F., Piper, S. C., and Heimann, M. (1996) Global and hemispheric CO_2 – sinks deduced from change in atmospheric O_2 concentration. *Nature*, **381**, 218–221.

Klimenko, V. V., Klimenko, A. V., Andreichenko, T. S., Dovganiuk, V. V., Mikushina, O. V., Konrad, S. G., Ivanova, G. A., and Fedorov, M. V. (1997) *Energy, Nature, and Climate* (215 pp.). Moscow: MEI Publishing [in Russian].

Koff, G. L., and Chesnokova, I. V. (1997) *On the Method of Insurance of Natural Risks: Analysis and Assessment of Natural and Technogenic Risk in Building* (pp. 128–130). Moscow: PNIIIS [in Russian].

Kondratyev, K. Ya. (1990) Key problems of global ecology. *Itogi Nauki i Tekhniki [Theoretical and General Problems of Geography]* (Vol. 9, 454 pp.). Moscow: VINITI [in Russian].

Kondratyev, K. Ya. (1992) *Global Climate* (359 pp.). St Petersburg; Nauka Publishing [in Russian].

Kondratyev, K. Ya. (1998) *Multidimensional Global Change* (771 pp.). Chichester: Wiley-Praxis.

Kondratyev, K. Ya. (1999a) *Ecodynamics and Geopolicy. Vol. 1: Global Problems* (1036 pp). St Petersburg: Russian Academy Science. Publishing [in Russian].

Kondratyev, K. Ya. (1999b) *Climatic Effects of Aerosols and Clouds* (264 pp.). Chichester, UK: Springer–Praxis.

Kondratyev, K. Ya. (2002) Global climate change: Facts, assumptions, and perspectives of research. *Optics of the Atmosphere and Ocean*, **15**(10), 1–15 [in Russian].

Kondratyev, K. Ya., and Cracknell, A. P. (1998) *Observing Global Climate Change* (592 pp.). London: Taylor & Francis.

Kondratyev, K. Ya. and Demirchian, K. S. (2001) Global climate change and Kyoto Protocol. *Herald Russian Acad. Sci.*, **N11**, 1–29 [in Russian].

Kondratyev, K. Ya. and Isidorov (2001) Global cycle of carbon. *Optics of Atmosphere and Ocean*, **14**(2), 89–105 [in Russian].

Kondratyev, K. Ya. and Losev, K. S. (2002) Present problems of the global civilization development and its possible perspectives. *Studying the Earth from Space*, **N2**, 1–21 [in Russian].

Kondratyev, K. Ya. and Varotsos, C. A. (2000) *Atmospheric Ozone Variability: Implications for Climate Change, Human Health, and Ecosystems* (758 pp.). Chichester, UK: Springer–Praxis.

Kondratyev, K. Ya. and Varotsos, C. A. (2003) *Global Carbon and Climate Change* (344 pp.). Chichester: Springer-Praxis.

Kondratyev, K. Ya., Demirchian, K. S., Baliunas, S., Adamenko, V. N., Boehmer-Christiansen, S., Idso, Sh. B., Kukla, G., Postmentier, E. S., and Soon, W. (2001a) *Global Climate Change: Conceptual Aspects* (125 pp.). St Petersburg: RFFI [in Russian].

Kondratyev, K. Ya., Losev, K. S., Ananicheva, M. D., and Chesnokova, I. V. (2001b) Elementary structural units of the biosphere and landscapes. *Doklady Acad. Sci.* **380**(1), 136–37 [in Russian].

Kondratyev, K. Ya., Losev, K. S., Ananicheva, M. D., and Chesnokova, I. V. (2001c) Some problems of landscape science and ecology in the context of the concept of biotic regulation. *Izv. Russ. Geogr. Soc.*, **5**, 22–29 [in Russian].

Kondratyev, K. Ya., Grigoryev, Al. A., and Varotsos, C. A. (2002a) *Environmental Disasters: Anthropogenic and Natural* (484 pp.). Chichester, UK: Springer–Praxis.

Kondratyev, K. Ya., Krapivin, V. F., and Phillips, G. W. (2002b) *Global Environmental Change: Monitoring and Modelling* (761 pp.). Heidelberg; Springer-Verlag.

Kondratyev, K. Ya., Losev, K. S., Ananicheva, M. D., and Chesnokova, I. V. (2003) The price of Russia's environmental services. *Vestnik RAN.*, **73**(1), 3–13 [in Russian].

Konrad, S. G. and Ivanova, G. A. (1998) Differentiated approach to the quantitative estimation of carbon emission in forest fires. *Forest Science*, **N3**, 28–35 [in Russian].

Kovda, V. A. (1990) *The State Ecological Policy of the Use and Protection of the Biosphere* (72 pp.). Pushchino [in Russian].

Krasilov, V. A. (1992) *Nature Protection: Principles, Problems, Priorities* (174 pp.). Moscow: Institute of Nature Protection [in Russian].

Landscape Protection. Explanatory Dictionary (1982) Moscow: Progress Publishing (272 pp.) [in Russian].

Lapo, A. V. (1987) *Traces of the Past Biospheres*. Moscow: Znanie Publishing [in Russian].

Losev, K. S. (1989) *Water* (272 pp.). Leningrad: Gidrometeoizdat [in Russian].

Losev, K. S. (2003) Natural-scientific basis of sustainable life. *Herald of Russian Acad. Sci.*, **73**(2), 110–16 [in Russian].

Losev, K. S. and Ananicheva, M. D. (2000) *Ecological Problems of Russia and Contiguous Territories* (283 pp.). Moscow: Noosphere Publishing [in Russian].

Losev, K. S., Gorshkov, V. G., Kondratyev, K. Ya., Kotliakov, V. M., Zalikhanov, M. Ch., Danilov-Danilyan, V. I., Golubev, G. N., Gavrilov, I. T., Reviakin, V. S. and Grakovich, V. F. (1993) *Problems of Ecology in Russia* (350 pp.). Moscow: VINITI Publishing [in Russian].

Losev, K. S., Ananicheva, M. D., and Chesnokova, I. V. (2001a) Landscape science and ecology – relationship and structural units. *Ukrainian Geography Journal*, **36**(4), 51–56 [in Russian].

Losev, K. S., Sadovnichy, V. A., Ushakova, I. S., and Ushakov, S. A. (2001b) *Biosphere and Humankind on the Way to Dialogue* (192 pp.). Moscow: Moscow State University Publishing [in Russian].

Lovelock, J. E. (1982) *Gaia. A New Look at Life on Earth* (157 pp.). New York: Oxford University Press.

Ludwig, W. (1997) Continental erosion and river transport of organic carbon to the World Ocean. *Sciences geologues*, **N98**, 1–196.

Macfedien, E. (1965) *Ecology of Animals (Goals and Methods)* (375 pp.). Moscow: Mir Publishing [in Russian].

Makaryeva, A. M. (2000) Biotically maintained stability of the Earth's mean global surface temperature. Preprint No. 2384, 42 pp. Russian Academy of Sciences, St Petersburg Nuclear Physics Institute.

Miagkov, S. M. (2001) *Social Ecology. Ecocultural Foundations of Sustainable Development* (190 pp.). Moscow: NIiPI of the Urban Ecology [in Russian].

Mohammed, E. O. and Abdel Rahman, B. A. (1998) Hazards in Africa: Trends, implication and regional distribution. *Disaster Prevention and Management*, **7**(2), 103–112.

Moiseyev, N. N. (1995) *Modern Rationalism* (376 pp.). Moscow: MGVP KOKS [in Russian].

Mokrousov, A. T. and Kudeyarov, V. N. (1997) Sources and sinks of carbon dioxide on the territory of Russia. In: *Global Environmental Change and Climate. Selected Scientific Works of the Russian Leading Scientists* (pp. 292–305). Moscow: Russian Academy of Science [in Russian].

Mooney, H. A. (1999) On the road to global ecology. *Ann. Rev. Energy. Environ.*, **N24**, 1–31.

Myneni, R. V., Dong, J., Tucker, C. J., Kaufmann, R. K., Kauppl, P. E., Liski, J., Zhou, L., Alekseyev, V., and Hughes, M. K. (2001) A large carbon sink in the woody biomass of northern forests. *Proc. Nat. Acad. Sci. USA*, **98**(26), 14784–14789.

National Interests and Safety Problems of Russia (108 pp). (1997) Moscow: Gorbachev Foundation [in Russian].

On the Environmental Condition of the Russian Federation for 1988–1998 (1999) Moscow: State Centre of Ecology (136 pp) [in Russian].

Osipov, V. I. (1995) Natural disasters in the centre of attention of scientists. *Vestnik RAN.*, **65**(6), 483–495 [in Russian].

Osipov, V. I. (1997) Natural disasters and sustainable development. *Geoecology*, **N2**, 5–18 [in Russian].

Our Common Future (374 pp.). (1989) Moscow: Progress Publishing [in Russian].

Parson, R. (1969) *Nature Presents a Bill* (568 pp.). Moscow: Progress Publishing [in Russian].

Pearson, P. N. and Palmer, M. P. (2000) Atmospheric carbon dioxide concentration over the past 60 million years. *Nature*, **404**, 695–699.

Porteous, A. (1996) *Dictionary of Environmental Science and Technology* (635 pp.). Chichester, UK: John Wiley & Sons.

Protection of Landscapes Glossary (1982) Moscow: Progress (272 pp.) [in Russian].

Protecting Tropical Forests: A High Priority Task (1990) Bonn: Bonner Universitätsbuch-druckerei (968 pp.).

Putin, V. V. (2000) Raw-material resources in the strategy of the economic development in Russia. In: *Russia in the Surrounding World 2000* (pp. 18–28). Moscow: MNEPU.

Reimers, N. F. (1990) *Nature Use* (640 pp.). Moscow: Mysl' Publishing [in Russian].

Reclus, J.-J. E. (1876) *Nouvelle Geographie Universelle. La Terra et les Hommes* (Vol. 1). Paris.

Reteium, A. Yu. and Serebriany, L. P. (1985) Geography in the system of Earth sciences. *Itogi Nauki i Tekhniki [Theoretical and General Problems of Geography]* (Vol. 4, 204 pp.). Moscow: VINITI [in Russian].

Review of the Global Economic and Social Situation 1996 (1996) New York: UN Economic and Social Council (515 pp.) [in Russian].

Rigor, I. G., Colony, R. L., and Martin, S. (2000) Variations in surface air temperature observations in the Arctic, 1979–1997. *J. Climate*, **13**(5), 896–914.

Rotmans, J. and Rothman, D. S. (eds) (2003) *Scaling in Integrated Assessment. Integrated Assessment Studies 2* (355 pp.). The Netherlands: Swets & Zeitlinger Publishing.

Rozenberg, E. S. (1999) Analysis of definitions of the notion "ecology". *Ecology*, **N2**, 89–98 [in Russian].

Russia in the Surrounding World, 2001 (2001) Moscow: MNEPU Publishing (332 pp.) [in Russian].

Science and Life (2002) Moscow: ANO "Nauka i Zhizn" (No. 1) [in Russian].

Serebrianny, L. R. (1980) *Ancient Glaciation and Life* (128 pp.). Moscow: Nauka Publishing [in Russian].

Severtsov, A. S. (1992) Dynamics of the size of population from the position of population ecology of animals. *Bull. Moscow Soc. of Nature Investigators, Biology,* **27**(6), 3–17 [in Russian].

Shelepov, V. V. (2000) Resource possibilities to provide the sustainable development of economy observing the ecological aspects of their use. In: *Ecological Aspects of the Energetic Strategy as a Factor of Sustainable Development* (pp. 35–42). Moscow: Noosphere [in Russian].

Shlichter, S. B. (1996) Contradictions of sustainable development and problems of their overcoming. In: *Geographical Problems of Sustainable Development of Nature and Society* (pp. 59–68). Moscow: Council on fundamental problems of geography [in Russian].

Schwarz, S. S. (1976) Evolution of the biosphere and ecological forecast. *Vestnik USSR Acad. Sci.,* **N2**, 61–72 (in Russian).

Smil, V. (1998) *Energies* (210 pp.). Cambridge, MA: MIT Press.

Sochava, V. B. (1970) *Geography and Ecology*. Leningrad: USSR Geographical Society Publishing.

Sochava, V. B. (1978) *Introduction to Science on Geosystems* (318 pp.). Novosibirsk, Russia: Nauka Publishing [in Russian].

Sorokhtin, O. G. and Ushakov, S. A. (1989) *Global Evolution of the Earth* (446 pp.). Moscow: MGU Publishing [in Russian].

Special double issue on the 21st century economy (1998) *Business Week*, 31 August.

State of the World, 1996 (1996) London: W. W. Norton Co. (249 pp.).

State of the World, 1999 (2000) Moscow: Ves Mir Publishing [in Russian].

Steffen, W. and Tyson, P. (eds) (2001) *Global Change and the Earth System: A Planet Under Pressure* (32 pp.). Stockholm: IGBP Science.

Sunkachev (1964) Principal definitions of forest biocenology. *Basis of Forest Biocenology*. Moscow: Nauka [in Russian].

Suess (1875) *Die Entstehung der Alpen*. Vienna [in German].

Tarko, A. M. (1994) A model of the global carbon cycle. *Nature,* **N7**, 27–32 [in Russian].

The Environmental Encyclopaedia Dictionary (1993) Moscow: Progress (640 pp.) [in Russian].

The Oxford Desk Dictionary and Thesaurus (1997) New York: Berkley Books (972 pp.).

Tickell, C. (1993) The human species: A suicidal success? *The Geographical Journal,* **159**(2), 219–226.

Timashev, I. E. (1999) *Geoecological Russian–English Dictionary* (168 pp.). Moscow: Muravey-Guide [in Russian].

Timofeev-Resovsky, N. V. (1998) *Biosphere and Humankind* (Scientific contributions Part 1). Obninsk, Russia: Obninsk Division of the USSR Geographic Society [in Russian].

Timofeev-Resovsky, N. V., Vorohtcov, N. N., and Yablokov, B. A. (1969) *Short Sketch of Evolution Theory*. Moscow: Nauka [in Russian].

Tinbergen, Ya. and Haavelmo, T. (1991) *Environmentally Sustainable Economic Development: Building on Brundtland* (100 pp.). Paris: UNESCO.

Tiuriukanov, A. N. and Federov, V. M. (1996) *Biospheric Thoughts* (368 pp.). Moscow: RAEN [in Russian].

Tol, R. S. J. (2003) Is the uncertainty about climate change too large for expected cost-benefit analysis? *Clim. Change,* **56**(3), 265–289.

Vallon, A. (1941) *Slavery History in the Ancient World*. Moscow: OGIZ-Gospolitizdat [in Russian].

Vavilov, N. I. (1926) Centres of cultivated plant origin. *Contributions to Applied Botany and Selection*, **16**(2).

Velichko, A. A. and Soffer, O. A. (eds) (1997) *Man Inhabits the Planet Earth* (304 pp.). Moscow: RAS [in Russian].

Vernadsky, V. I. (1944) Some words about noosphere. *Success of Recent Biology*, **18**(2), 118–120 [in Russian].

Vernadsky, V. I. (1967) *Biosphere* (376 pp.). Moscow: Mysl' Publishing [in Russian].

Vishnevsky, A. G. (1998) *Sickle and Rouble* (432 pp.). Moscow: OGI Publishing [in Russian].

Vital Signs 2001–2002. The Trends that are Shaping Our Future (251 pp.). London: Earthscan.

Vitousek, P. M., Mooney, H. A., Lubchenco, J., and Melillo, J. M. (1997) Human domination of Earth's ecosystems. *Science*, **N5325**, 494–499.

Wallace, A. R. (1878) *Natural Selection*. St Petersburg [in Russian].

Wallon, A. (1941) *History of Slavery in the Ancient World* (664 pp.). Moscow: OGIZ [in Russian].

Webb, W. P. (1969) *The Great Frontier*. Austin, TX.

Weber, A. B. (1999) *Sustainable Development as a Social Problem* (122 pp.). Moscow: RAS [in Russian].

Weitzsäcker, E. von, Lovins, A. B., Lovins, L. H. (1997) *Factor Four. Doubling Wealth – Having Resource Use* (322 pp.). London: Earthscan.

World Resources 1989–1889 (1989) New York: Oxford Basic Books (pp. xii, 372).

Index

Printing: Mercedes-Druck, Berlin
Binding: Stein+Lehmann, Berlin